Materials Science and Technologies

Waste and Waste Management

Materials Science and Technologies

Magnesium Alloys: Advances in Research and Applications
Catalin Iulian Pruncu, PhD (Editor)
Kavian Omar Cooke, PhD (Editor)
2022. ISBN: 978-1-68507-975-8 (Softcover)
2022. ISBN: 979-8-88697-237-5 (eBook)

Photosensitizers and Their Applications
Davor Margetic, PhD (Editor)
Renjith Thomas, PhD (Editor)
2022. ISBN: 978-1-68507-880-5 (Hardcover)
2022. ISBN: 979-8-88697-039-5 (eBook)

Evaluation of Reserves and Resources in Unconventional Reservoirs
Cenk Temizel (Author)
Cengiz Yegin, PhD (Author)
Shah Kabir (Author)
2022. ISBN: 978-1-68507-841-6 (eBook)

Aerogels: Properties and Applications
Alina Iuliana Pruna, PhD (Editor)
2022. ISBN: 978-1-68507-788-4 (Hardcover)
2022. ISBN: 978-1-68507-895-9 (eBook)

High-Performance Calcium-Carbonate Concrete
Natt Makul (Author)
2022. ISBN: 978-1-68507-412-8 (eBook)

Analytical Models of Interstitial-Atom-Induced Stresses in Isotropic Metallic Materials
Ladislav Ceniga (Editor)
2022. ISBN: 978-1-68507-429-6 (eBook)

More information about this series can be found at
https://novapublishers.com/product-category/series/materials-science-and-technologies/

Rajesh Kumar Verma
Jogendra Kumar
and Shivi Kesarwani

Recycling of Discarded Carpets for Structural Polymer Composites

Copyright © 2023 by Nova Science Publishers, Inc.
https://doi.org/10.52305/VNCP7234

All rights reserved. No part of this book may be reproduced, stored in a retrieval system or transmitted in any form or by any means: electronic, electrostatic, magnetic, tape, mechanical photocopying, recording or otherwise without the written permission of the Publisher.

We have partnered with Copyright Clearance Center to make it easy for you to obtain permissions to reuse content from this publication. Simply navigate to this publication's page on Nova's website and locate the "Get Permission" button below the title description. This button is linked directly to the title's permission page on copyright.com. Alternatively, you can visit copyright.com and search by title, ISBN, or ISSN.

For further questions about using the service on copyright.com, please contact:
Copyright Clearance Center
Phone: +1-(978) 750-8400   Fax: +1-(978) 750-4470   E-mail: info@copyright.com

NOTICE TO THE READER

The Publisher has taken reasonable care in the preparation of this book, but makes no expressed or implied warranty of any kind and assumes no responsibility for any errors or omissions. No liability is assumed for incidental or consequential damages in connection with or arising out of information contained in this book. The Publisher shall not be liable for any special, consequential, or exemplary damages resulting, in whole or in part, from the readers' use of, or reliance upon, this material. Any parts of this book based on government reports are so indicated and copyright is claimed for those parts to the extent applicable to compilations of such works.

Independent verification should be sought for any data, advice or recommendations contained in this book. In addition, no responsibility is assumed by the Publisher for any injury and/or damage to persons or property arising from any methods, products, instructions, ideas or otherwise contained in this publication.

This publication is designed to provide accurate and authoritative information with regard to the subject matter covered herein. It is sold with the clear understanding that the Publisher is not engaged in rendering legal or any other professional services. If legal or any other expert assistance is required, the services of a competent person should be sought. FROM A DECLARATION OF PARTICIPANTS JOINTLY ADOPTED BY A COMMITTEE OF THE AMERICAN BAR ASSOCIATION AND A COMMITTEE OF PUBLISHERS.

Additional color graphics may be available in the e-book version of this book.

Library of Congress Cataloging-in-Publication Data

ISBN: 979-8-88697-538-3

Published by Nova Science Publishers, Inc. † New York

Contents

Preface vii

Acknowledgments ix

Abbreviations xi

Chapter 1   Introduction 1

Chapter 2   Novel Methodology in the Fabrication of Composite from Carpet Waste 21

Chapter 3   Mechanical Properties of Composites Developed from Carpet Waste 45

Chapter 4   Physical and Morphology Properties of Composites Developed from Carpet Waste 71

Chapter 5   Applications of Discarded Carpets Composites 91

Appendix 1 109

Index 111

About the Authors 115

Preface

Nowadays, the manufacturing sector is generating waste in very bulky sizes, in which carpet and textile industries are producing tons of waste and discarded materials during the development of final products. Most of the textile sector waste is non-biodegradable due to the constituents of polymeric materials. Solid waste management is an innovative way to reuse the byproducts in developing new products for other applications, as it will save the cost issues and other environmental challenges. In the reutilization of waste, there are several critical issues for controlling many types of waste, including solid, liquid, and gaseous forms. The manufacturing of carpets and their uses is a substantial source of waste. In this context, the many techniques employed for reusing carpet waste products. The generation of industrial waste leads to the disturbance of ecology and environmental conditions by generating hazardous effects on human health. The principle of *"waste to wealth"* could be helpful for manufacturing industries and human beings. Due to their bulky sizes and decomposition cost, the waste generated from the carpet and textile industries is a matter of concern. Exploration to produce discarded carpet polymer materials is in its early stages and requires greater interest from academia, research, manufacturing, and organizations. This book deals with the reuse of textile sector waste for developing polymer products. Several innovative ways are presented to highlight the effects of carpet waste and its reuse for structural components fabrication. Systematically, this book is divided into different sections to understand the basics of carpet waste generation to the finished products.

Accordingly, Chapter 1 highlights the fundamental introduction to the carpet sector and waste generated from carpet and textile products. A novel approach is presented to utilize carpet waste and discarded products during carpet manufacturing to create lightweight structural applications. Chapter 2 highlights the introduction of Polymer studies with the properties, polymer-based composite materials, and their classification. It will provide a novel direction for reusing waste carpets into an appropriate hybrid form and provide knowledge on the methodology of mechanical characteristics and performance evaluation of developed composites. It would highlight the importance of the proposed book and research orientation. It provides the requisite knowledge of recycled waste carpet composite properties and scientific implementation. This method will reduce the discarded carpet and a simultaneously environmentally friendly approach shall be helpful for the manufacturing sector. Also, describe the overview of the limitations of the existing manufacturing methods by using the resin transfer molding method in a vacuum environment. Chapter 3, which focuses on carpet waste/epoxy composite analysis setup. It is primarily aiming mechanical characteristics (tensile, flexural and impact strength) based on prior research investigation and finding outcomes of waste carpet composites. Later, Chapter 4 explores the physio-thermal, mechanical and morphology (water absorptivity, fire retardancy, temperature) of developed carpet waste polymer composites. In this chapter, carpet waste/epoxy composite analysis setup has been discussed. Chapter 5 highlights the essence of finding some solution to convert such solid waste into a functional form with several applications. This chapter highlights the fundamental development product from recycled carpet and textile products.

Acknowledgments

This book entitled, *Recycling of Discarded Carpets For Structural Polymer Composites*, is possible due to the help, support and direction of several eminent people and scholars. First, my most profound gratitude goes to the Almighty, who endowed me with health, grace, wisdom, patience, and direction to complete this work. My most profound appreciation goes to Prof. Samsher, Hon'ble Vice Chancellor, HBTU Kanpur, for creating a research-oriented environment in the university and continuous direction for the timely completion of this assignment. Under his exemplary visionary leadership and infrastructural facility support, this work completion is possible on time. I am extremely grateful to him for motivating me to think independently, good-naturedly and solving numerous problems with perseverance. We are very thankful to Prof. J.P. Pandey, Hon'ble Vice Chancellor, MMMUT Gorakhpur, for their encouragement and enthusiastic support during this bookwork. I am fortunate to have their advice and support to express my thankfulness for their support and belief.

I am extremely grateful to the University Chancellor and Hon'ble Governor of Uttar Pradesh and the Government of Uttar Pradesh state for providing us with world class infrastructure and other technical support to the university which helped for completing this task. I want to express special thanks to the Additional Development Commissioner (Handicrafts), under Ministry of Textiles, Govt. of India, for their continuous support and direction in writing this book.

The authors would like to express their gratitude to Prof. Balvir S. Tomar of NIMS University Jaipur, India, for their generous support. We are highly appreciative and grateful to the Indian Institute of Technology, Roorkee (Saharanpur Campus), Central Institute of Petrochemicals Engineering and Technology (CIPET) Lucknow, Babasaheb Bhimrao Ambedkar University (BBAU) Lucknow for their constant laboratory support in completing the current work. The authors are very indebted for providing technical information and support from Prof. Sanjay Mishra, Mr. Balram Jaiswal, Mr.

Rahul Vishwakarma, Mr. Kuldeep Kumar and Mr. Kaushlendra Kumar of the Materials and Morphology Laboratory group. Also, for their valuable asset and for providing the expertise information to complete this work. We are very gratified to Prof. K.K. Goswami, Dr. S.K. Gupta, IICT Bhadohi, India and all the faculty/staff members. This work is impossible without the blessings and kind support of my global research collaborators from various reputed institutes.

The authors are very gratified to the reviewers, editorial members, and NOVA publication staff for their valuable time and effort in this book assignment. Their efforts were significant in producing this book in its present form. We could not have accomplished this milestone without their constant and consistent advice, support, and collaboration.

Prof. Dr. Rajesh Kumar Verma
Lead Author

Abbreviations

CDF	Carpet-Derived Fuel
SiC	Silicon Carbide
Al_2O_3	Aluminium Oxide
FRP	Fiber Reinforcement Polymer
MMC	Metal Matrix Composite
CMC	Ceramic Matrix Composite
PMC	Polymer Matrix Composite
LDPE	Low-Density Polyethylene
MAh	Maleic anhydride
rPET	Recycled Polyethylene Terephthalate
SEM	Spectroscopy Emission Method
XRD	X-Ray Diffraction
ILSS	Interlaminar Shear Strength
IFSS	Interlaminar Flexural Shear Strength
FBBF	Front Back-Back Front
BFFB	Back Front-Front Back
NFC	Nylon fiber-based composites
OFC	Olefin fiber-based composite
PVA	Polyvinyl Alcohol
GFRP	Glass Fiber Reinforcement Polymer
POFA	Palm Oil fuel Ash
PVC	Poly Vinyl chloride
VARTM	Vacuum Assisted Resin Transfer Moulding
FTIR	Fourier-Transform Infrared Spectroscopy
G/CFRP	Graphene/Carbon Fiber Reinforcement Polymer
PP	Polypropylene
R	Resin
H	Hardener
NRC	Nylon Reinforced Composites
RWCA	Recycled Waste Ceramic Aggregate

WCF	Waste Carpet fiber
HSC	High Strength Concrete
NCA	Natural Coarse Aggregate
ASR	Automotive Shredder Residue
NFRC	Nylon fiber Reinforced Concrete
NFFRC	Nylon fiber Fabric Reinforced concrete
UCS	Unconfined Compressive Strength
SWF	Ship Wool Fiber
LWA	Light Weight Aggregates
HC	Hessian Cloth
CBC	Carpet Backing Cloth
GNP/MWCNT	Graphene Nanoplate/Multiwall Carbon Nanotube
ASTM	American Standard Testing Materials
UTM	Universal Testing Machine
FA	Fly Ash
TGA	Thermogravimetric Analysis
FRC	Fiber Reinforced Concrete

Chapter 1

Introduction

Waste has become the prime concern for the industrial and manufacturing sectors in the 21st century. Waste management is important in managing different types of waste, including solid, liquid, and gaseous forms. Industries, businesses, research organizations, and academe institutions use numerous techniques to control waste and resource development (Wilson 2007). Carpet manufacturing and its commercial and domestic use are significant sources of waste. In this context, there are many techniques used to recycle carpet waste. In the manufacturing sector, waste gets generated from the initial product development to the end products. This leads to an imbalance of the environmental conditions and hazardous effects on the ecology. Proper collection and recycling of discarded products is a viable waste management solution (Islam and Bhat 2019; Siddique, Khatib, and Kaur 2008). The *"Waste to Wealth"* principle can benefit the manufacturing sector and people's economic issues. Waste generated in the carpet and textile industry is a concern due to its bulky size and decomposition cost. Research into the production of polymeric waste carpet materials is still in its infancy and needs more interest from academia, research centers, manufacturers and professional societies. This book provides a basic introduction to the carpet industry and the waste generated from carpets and textiles. A new approach is presented to create lightweight structural applications using carpet waste. An attempt has been made to overcome the limitations of the existing manufacturing processes by using the Resin transfer molding method in a vacuum environment. This book presents a new direction for recycling carpet waste in a suitable composite form. The method is said to reduce carpet waste and the eco-friendly approach will benefit the converting industry.

History

The word carpet is derived from the old French "CARPITE," According to the dictionary, it first appeared in the late 13th century and meant "Coarse Cloth". It came to mean "Tablecloth or bedspread" until the 14th century. Sometimes, the term "rug" can be used instead of "carpet". But the rug is generally considered to be smaller as compared to carpet. Usually, the carpet was prepared from wool before the 20th century. Still, after the 20th century, artificial fibers such as nylon, polyester, and polypropylene are commonly used because they are less expensive and have greater design flexibility than natural fibers (Kalebek 2016).

The carpet industry in India is one of the oldest and most prevalent sectors. India has a long history of carpet weaving, a technique that draws on skills and expertise from worldwide, including Persia, China, and Afghanistan. Mughal Emperor Babar was displeased with the lack of amenities in India as compared to Persian luxuries, including carpets. They developed a carpet manufacturing industry (Agra) for the Persian carpet in India from 1580 onwards. Later, to speed up the production of Persian-styled carpets, Akbar established carpet-weaving enterprises in Agra, Delhi, and Lahore. The main centers of the carpet industry experienced rapid growth in India's north-western region (Kashmir, Jaipur, Agra, Bhadohi, and Mirzapur) (Bano and Fatima 2014). Between the 1930s and 1990s, almost all carpets were made with synthetically colored wool. An embargo on Iran diverts significant exports to China, India, and Tibet in the 1980s. The use of organically colored wool in hand-knotted oriental rugs became popular in the 1990s. Nowadays, carpets are a three-dimensional comfort product to decorate houses, offices, and commercial places. Carpet industries take several folds of advancements using different fibers, backing materials, handmade & machine-made fabrication methods, different colors, shades, designs and thicknesses as per the need and comfort of users.

Historical Hierarchy of Advancement in Carpets

There are changes in the textile era and the development of various carpets with unique designs. This section describes the phases and history of carpet development and its advancements. The general areas where carpets and rugs may first have been developed are thought to be Persia (Iran), China, Turkmenistan, Central Asia, and Mongolia, and it is often referred to as the carpet belt. It is said that the tomb of the Persian King Cyrus, who was concealed at Pasargadae (Persepolis), was covered by prized carpets. Even before his time, it is very likely that Persian wanderers identified using knotted carpets. In the provinces of Azerbaijan and Hamadan, where the Seljuk effect was most potent and longest-lasting, the Turkish knot is still used to this day.

Ornamental rugs or carpets have continually played an essential role in Islamic culture in specific and attained unparalleled heights in Baghdad, Damascus, Cordova, Delhi, and the famous cities of Central Asia. The use of carpet in Arabic and Persian literature are numerous in different traditional activities. On the basis of chronological viewpoint, the carpets are categorized into three alternates, starting with *Antique carpets*, which have been weaved for longer than a century (before the 1920s). These are traditional woven carpets with natural dyes before introducing synthetic dyes (during the 1860s-1870s). Next comes in the historical timeline of carpets are *Semi-antique carpets*, and these were made in the early to the middle part of the 18th century, from around the 1800s up to about the 1900s. In this era, the carpet industry saw a revamp. The rise of new, less costly carpets did away with the need for hand-woven carpets in the market following the economic depression and the short-lived post-depression market period. Finally, the modern age of carpets is defined as *Modern carpets* which are woven from the 1900s to today and are recognized as modern carpets. Western preferences and needs are generally accepted to have affected these rugs. Carpets woven before the 17th century are extremely rare and can only be found in Encyclopedic and Historical Museum collections. Figure 1.1 summarizes the prime phases of advancement in carpet manufacturing.

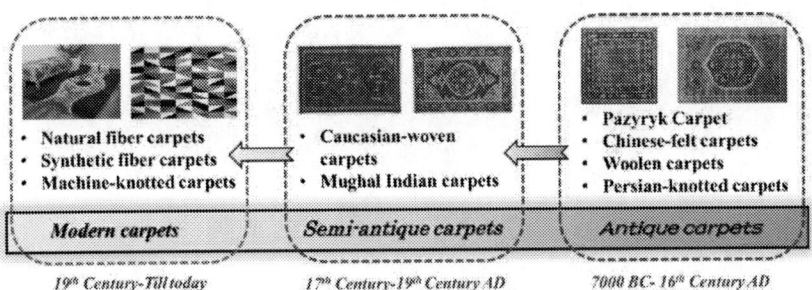

Figure 1.1. Different phases in carpet advancements.

- *7000 BC-540 BC (Neolithic age):* The use of warp/weft textiles in the flat weave kilims and also the Egyptian fresco of handloom (discovered in 1953). The initial pile carpets developed in the world are termed 'Pazyryk Carpet' (Goswami 2009b, 2018).
- *3rd Century:* Indian woolen carpets
- *Before 6th Century:* Persian-knotted and Turkoman (Turkmenistan)-woven carpets

- *8th Century:* Caucasian-woven and Chinese-felt carpets
- *12th Century:* Turkish-knotted carpets
- *16th Century:* Mughal Indian carpets

Major Carpet Belt Area

Carpets have been utilized for various purposes, including tile or concrete floor insulating, crafting a more decorative and comfortable floor for sitting, and sound insulation. The carpet industry has flourished with different designs, fabric materials, colors, shades, thicknesses, etc., at a worldwide level (Afghanistan, Armenia, Azerbaijan, China, India, Pakistan, Persia, Iran, Scandinavia, Turkey, Uzbekistan, Spain, Serbia, Bulgaria, France, and England) have a significant area (as displayed in Figure 1.2). Carpet weaving has a long history that draws on experience and skillsets from all over the world, including Persia, China, and Afghanistan, to establish a beautiful and functional technique. Furthermore, Abhushan and Turkman designs with beautiful flower borders surround elegant and simple motifs.

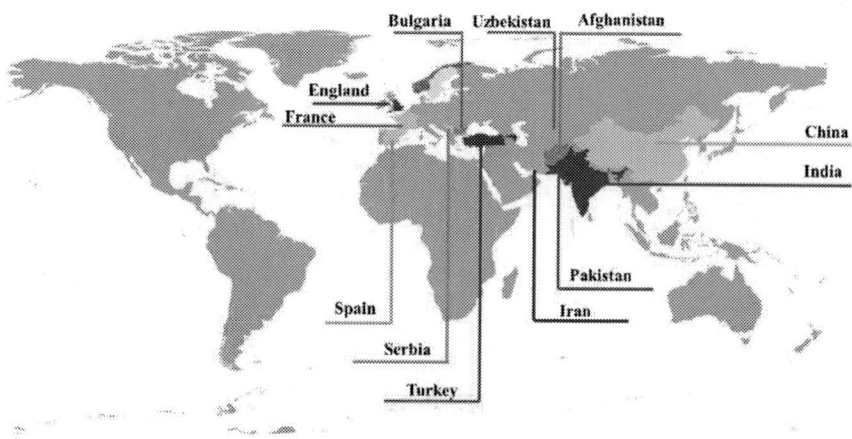

Figure 1.2. Major carpet belts at worldwide level.

Jaipur and Bikaner in Rajasthan (A state in India), Bhadohi, Mirzapur, Agra, Kashmir, and their border areas are the primary industrial and manufacturing belt in the carpet and textiles sector. Uttar Pradesh (A state in India) carpets are known worldwide for their distinctive colors and designs. Bhadohi, Mirzapur, and Agra are the three main carpet centers in Uttar

Pradesh. Since Agra (Uttar Pradesh) was the capital of Akbar's kingdom, it was the first carpet center in Mughal India. This city is well-known for its Persian-style carpets. Kashmir is the epicenter of Indian oriental carpets and rugs, given their popularity worldwide. Hand-knotted woolen and silk carpets are known for their quality and artistic expertise. India's carpet-producing centers are in Kashmir, and these are famed for their traditional hand-knotted silk carpet weavers. Kashmir's conventional silk carpets can be traced back to the Mughal Empire. Kashmir's carpet's look and design are based on a Persian carpet layout and pattern. Rajasthani carpets have been long noted for their high-quality hand-knotted woolen fibers. Jaipur, Ajmer, and Bikaner are the prominent locations for this craft in Rajasthan. During the Mughal Empire, Jaipur was another major carpet weaving center. Artists from Afghanistan arrived in Rajasthan's royal ateliers in the 17th century and flourished carpet production in Rajasthan.

Carpet Manufacturing Technique

There are various types of carpets available in India. Every type of carpet in India has a unique design that attracts different spectators and buyers. Carpet weaving utilizes multiple techniques, including some Mughal period procedures, which are still in use today. Some specific methods have been outlined here for better understanding as follows.

Hand Made

The Mughals pioneered and initiated the handmade technique and created different designs on fabrics. This carpet has economic potential because workers hired by the ruling elite made it. Initially, the patterns were Turkish and Persian in origin but were eventually Indianized. Kashmiri art incorporates many techniques and designs. In this method, a backing material is pumped into the carpet pile using a tufting gun and then attached to a secondary backing fabric with a latex solution for stability. The design is then completed with the usage of the third backdrop cloth. Tufted carpets can be made with either a loop pile or a cut pile. It is a combination of hand-knotted carpets and flatweave carpets. Compared to walking on a bare concrete floor, rugs can be utilized for various commercial and domestic purposes, including room decoration, a comfortable location to sit, walk, etc., and sound absorption. They have a lot of design flexibility in terms of fiber morphologies, color, and fiber length.

Machine-Made

The machines and automatic devices are widely used in the development of carpets with unique designs and dimensions. The most common types of machine-made carpets include tufted, woven, knitted, flocked, needle-punched, needle felt and braiding, fusion bonding, etc. In general, the tufting mechanism has primarily been used the preparation of machine-made carpets and rugs. The tufted method uses the automated setup and software in which the yarn is stitched through a preconstructed backing to form a loop or a tuft. To clutch the loops in a specific place, the back portion of the carpet backing is coated using latex material. The tufting mechanism is a very cost-effective, durable and time-consuming method for fabricating carpets and rugs through machines. Carpet artisans can efficiently regulate the different machine and process parameters. These parameters are highly responsible for quality and productivity indices such as length, thickness, colors, patterns, stitching rate, surface shades, backing materials, etc.

Recently, various tools and optimization techniques have been used to save time and cost issues in the development of machine-made carpets (Verma, Dwivedi, and Sinha 2017). The findings of optimization modules show a significant improvement in the quality functions of the machine-made carpets. This also enhances durability and controls the cost-effectiveness issues during manufacturing phases.

Using a vacuum-based mechanism for yarn filling during carpet manufacturing boosts the efficiency of the entire process. The carpet is then rolled into length, packaged in solid plastic sheets, and shipped to either the manufacturer's inventory warehouse or a retail carpet store. Over 95% of the carpet mass-produced in the United States is tufted type (Binggeli 2011). A tufting machine involves an automatic mechanism of sewing with the help of several needles and punches. It inserts loops of yarn into the primary backing sheets. The tufting mechanism can produce different loops, such as a level loop, multilevel loop, cut pile, and cut-and-loop pile structures (Goswami 2018, 2009b; Mohapatra and Verma 2017).

Classification of Carpet

The most commonly used carpet is available in the small, medium, and mass-scale production sectors (ref. Figure 1.3).

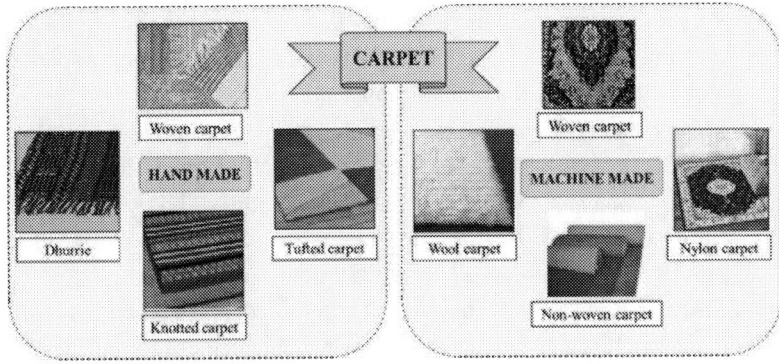

Figure 1.3. Classification of carpet.

- *Tufted Carpet:* The tufted carpet is still manufactured today on a mass scale for domestic and commercial purposes. Tufting is the stitching of face yarns onto a backing material with multi-purpose machinery. The fibers are secured to the pre-woven backing by a thick latex coating. An additional backing can be employed for added dimensional stability.
- *Woven Carpets:* These carpets are woven on large looms in the same way any other cloth is woven. They could have loops and stacks that are cut and uncut. A range of bright yarns on the carpets is utilized to create various patterns and designs.
- *Non-Woven Carpets:* Non-woven carpets are made in a different way than tufted and woven carpets. Non-woven carpets are made from polypropylene fibers. Needle punching creates a web from the strands, which is then thermally fused to create the rug. They are most used in automotive applications and for short-term uses such as exposition halls and foot-mats.
- *Flatweave Carpets:* Flatware carpet is used as a floor covering in a specific location. Interlocking warp and weft threads are used to achieve this significance. Oriental flatwoven carpets include kilim, soumak, plain weave, and tapestry weave. This is very common in Asian countries, such as Japan.

- *Needle-Felt Carpets:* The needle-felt carpet is significantly better in terms of technology. Individual fibers are attracted to one another by electrostatic attraction, resulting in a one-of-a-kind carpet with exceptional durability. A needle felt mat is used in commercial and industrial applications such as hotels and high-traffic areas.
- *Velvet Carpets:* Velvet rugs/carpets are incredibly soft and have a uniform color; they are delicately twisted, and their rich appearance makes them ideal for formal rooms.
- *Hand-woven Carpets:* Hand-knotted and hand-woven carpets and rugs are treated in the same way. On the other hand, floor coverings that are hand-woven or knotted differ slightly. There is no pile of flatweaves. They are woven on a particular sort of loom with very precise weft and warp placement.
- *Fiber Type Carpets:* The type of fiber used to make carpets affects their durability, appearance, and cost. The most common fiber kinds include nylon, olefin, polyester, acrylic, wool, and blends.
- *Nylon Carpets:* The nylon type is the most common fiber used in carpets. Nylon is a durable, long-lasting material that can withstand environmental factors. When exposed to direct sunlight for an extended period, it develops a static electricity conductor that fades.
- *Olefin Carpets:* The olefin carpet is a modern carpet. These carpets are ideal for outdoor carpeting since it is mold and mildew-resistant.
- *Polyester Carpets:* Polyester is gaining popularity as a less expensive alternative to other textiles. It has a smooth feel when combined with a thick cut-pile construction. Because it is less resilient and prone to breakage and fading than nylon fiber, it is unsuitable for high-traffic areas.
- *Acrylic Carpets:* Acrylic carpet is not a popular material, but it mimics the look and feel of wool-based carpets for a fraction of the price. It is mold and mildew-resistant but does not have a higher electrical conductor carrying capacity.
- *Wool Carpets:* Wool carpet is the only natural fiber used in carpet production and the most expensive carpet material available today. It provides a lovely feel against bare feet and is quite long-lasting. It is stain-resistant, but it fades swiftly when exposed to direct sunlight.
- *Blends Carpets:* Blend material carpets with these fiber blends has a better overall look and feel and more excellent durability. Wool and nylon are the most prevalent combinations.

Application of Carpet

Carpets for household use are produced all over the world, as it holds a significant market capital in necessary household utility items. China and the Indian subcontinent are becoming huge carpet-producing locations. The different types of carpets are classified in Figure 1.4, and their feasible applications are described in detail. Carpets are excellent thermal and sound insulators in addition to decorative commodities. The insulating and other required aesthetic properties of carpet with underlay can effectively lessen heat loss through the floor when used with traditional heating systems. Carpets can have a thermal insulation rating of 0.1 m^2K/W to 0.3 m^2K/W, depending on the construction and specification. It is assumed that when underfloor heating is used, carpets act as a heat transmitters to the airspace above. Carpets can provide acoustic insulation in three different ways. These are material characteristics that have an impact on sound and acoustic absorption. This is about the sound that is transmitted into the chamber below, such as footsteps or a dropped object. The carpet pile absorbs impact energy, converting a high-frequency thud to a low-frequency thud with far less ear damage.

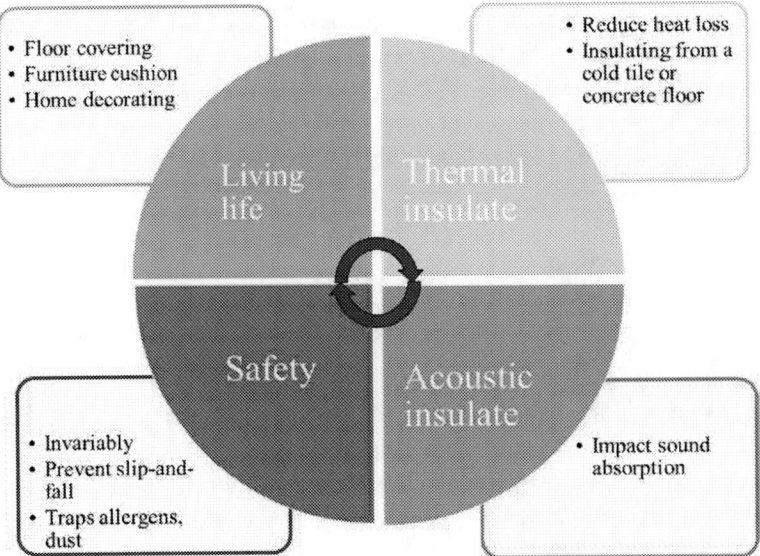

Figure 1.4. Application of carpet.

Overview of Discarded Carpets: Waste Generation Issue

Nowadays, carpet waste is a considerable concern for the environment and the economy due to the costs involved in the deposal of waste carpets. Only a small portion of it gets recycled yearly compared to carpet waste generated globally through different sources (Goswami 2009a; Jain et al. 2012). Carpets are widely used in the form of decorative applications and for comfort objectives as floor covering, commercial offices and buildings, etc. In our daily lives, almost all of us see carpets in different forms around us. Since India is the world's second-largest country after China, with a population of around 130 crores it generates a large amount of carpet waste every year (Henckens 2021). Carpet waste of about 40,000 tons is sent to a landfill in the UK itself (Mohammadhosseini et al. 2018). Due to this, it has become a global concern. Only a small percentage of the amount of carpet waste produced yearly gets recycled. Carpet waste recycling requires multiple processing steps, making it difficult and expensive (Realff, Ammons, and Newton 1999). Considering carpet waste as landfilling decisions become uneconomical due to its non-biodegradable behavior and high cost of landfilling. It is a type of solid waste that is causing environmental issues. Thus, it becomes essential to find some solution to convert such solid waste into a useful form with several applications. Burning this fibrous waste releases highly toxic fumes, which harm human health (Mishra et al. 2019; Ghobakhloo and Fathi 2021; Junxiong Wang et al. 2021). Carpet waste can be collected in two forms, depending on its source of origin (Ahmed et al. 2021; Teli 2018).

Pre-Consumer Waste

Pre-consumer carpet waste includes the trimmed scrap generated at the time of manufacturing or production of carpets per customer needs and demands, such as in movie theaters, households, and educational or institutional complexes. Figure 1.5 demonstrates the trimmed carpet waste collected from manufacturing firms or factories. These are collected from garbage piles or dustbins. Pre-consumer carpet waste includes the trimmed scrap generated during the manufacturing or production of carpets per customer needs and demands, such as in movie theaters, households, and educational or

Introduction

institutional complexes (Jaiswal et al. 2021; Kumar, Mishra, and Verma 2019). To make a suitable design according to the requirement, the carpet is trimmed out, and this trimmed part cannot be utilized further at the same place. This generated waste is shrunken and irregular and cannot be used further directly. About 10% of the total carpet waste is generated as scrap (Miraftab 2018; Y. Wang, Ucar, and Wang 2014).

Post-Consumer Waste

Post-consumer waste comprises carpet waste generated after it becomes useless in households and commercial buildings (Sotayo, Green, and Turvey 2015; Dhawan et al. 2019; Cline and Friddle 1992). After consumers have used it up to a certain period, the waste carpet is thrown into dustbins. Figure 1.5 depicts the waste carpet after it has been used by consumers and thrown into dustbins. The users directly throw these carpets as waste. The carpet waste management hierarchy is depicted in Figure 1.5. A summarized overview of the waste generated from carpets, as investigated in various studies, has been described in Table 1.1.

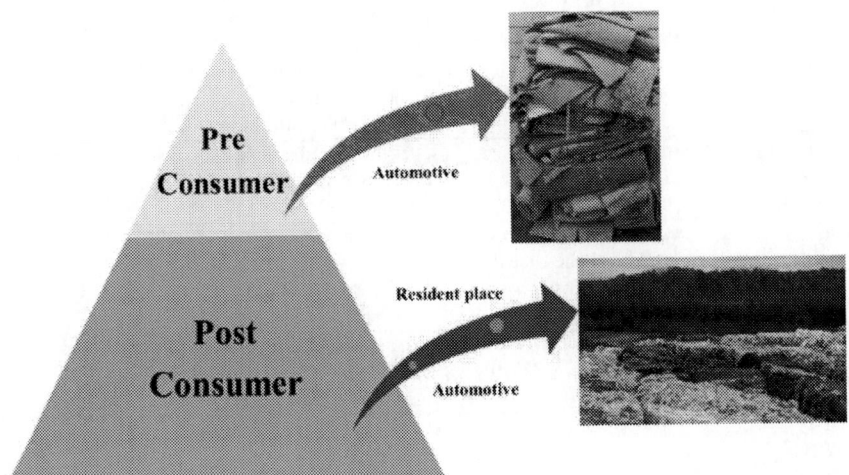

Figure 1.5. Types of carpet waste.

Table 1.1. Summary of produced carpet waste

Sr. No.	Types of Carpet waste	Country	Quantity (Tons)	Ref.
1	Carpets are sent to landfills in the UK annually post-consumer	US	400000 tonnes	(Sotayo, Green, and Turvey 2015)
2	Carpet made of polypropylene	Every year, fibers are dumped in Malaysian landfills. (2018)	70 million tonnes of garbage per year and 50 tonnes of industrial waste 4	(Mohammadhosseini et al. 2018)
3	Municipalities, recycling companies, and carpet trash generated by the residential, commercial, and industrial sectors.	In the United States of America, around	According to reports, 500,000 tonnes of carpet waste fiber are dumped into landfills in the United Kingdom each year, and 10 million tonnes of textile trash were created in 2003.	(Mirzababaei et al. 2013)
4	Solid waste management in India	India	Every year, 62 million tonnes of garbage are generated. Approximately 43 million tonnes (70 percent) are collected, with approximately 12 million tonnes being processed and 31 million tonnes being disposed on landfill sites.	(Sahana 2021)
5	Post-Consumer Carpet Composite	USA	Every year, 250 million tonnes of carpet garbage are created.	(Mishra et al. 2019)
6	Every year, post-consumer carpet trash is dumped in landfills.	Currently, in the United States alone (2007)	over 2 million tonnes	(Y. Wang 2007)
7	Organic materials as like rubber, leather, and textile fibers.	MSW in the U.S. (2010)	13.12 million tonnes	(Drescher et al. 2014)

Carpet Waste Recycling Methods

Green manufacturing, waste management, application, and environmental development are the most important things to keep in mind in the industrial sector in the modern era. With the generation and increase of carpet waste from many fields, its re-utilization has become a matter of concern to convert this waste into a valuable product to maintain waste material (Cunningham, Green, and Miller 2021; Liu et al. 2021; Chen et al. 2021). Numerous studies and experimental investigations were made by eminent scholars, researchers and manufacturing organizations to control waste generation and their re-utilization for product development. It is required to reuse the generated waste in the interest of society and the environment. In a worldwide level, several institutions, government sector departments and non-government organizations (NGOs) are trying to control waste issues. They are working on solid waste management principles to achieve the "waste to wealth" objective. Various recycling processes have been proposed based on the methodology applied and product quality production (Table 1.2). The following categories have been presented for carpet waste recycling:

Primary Recycling

Primary recycling, sometimes referred to as "closed-loop recycling," is the process by which recyclables are systematically processed to generate a product that fulfills a function that is analogous to the intended use of the recyclable materials. This involves detaching the face fibers from their backing. This can be done manually or by a particular designed machine. When materials are recycled several times, their quality deteriorates, especially in polymers and textile products. Weaved strands tangle together to give textile items their strength. Because of the constant recycling of textiles, the fibers get shorter and shorter. Eventually, they are no longer used in any other kind of recycling. Similarly, each time polymeric materials are recycled, the quality of the materials diminishes, and it is believed that these materials may only be recycled one to two times before quality deteriorates.

Secondary Recycling

Secondary recycling is similar to primary recycling in that it is also a mechanical recycling process; however, it employs recycled materials in the production of brand-new items. This new product often does not have the same physical needs as the original product and is typically less recyclable than the original product. Recyclability has decreased, yet this is still a significant step in reducing the use of virgin resources and diverting fewer recyclable products like waste materials from landfills. In terms of polymeric waste, secondary recycling requires recouping the individual parts of a polymeric mixture without necessarily breaking them down to the monomeric form. This class comprises several extractions processes.

Tertiary Recycling

Tertiary recycling or melt blending is a physical methodology. It is a widespread procedure due to its flexibility and simplicity (Xanthos et al. 2002; Mohammadhosseini, Awal, and Mohd 2017). It is the process by which polymeric and textile-based materials are transformed into petrochemical and fuel feedstock. Due to the high energy demand and high cost of primary and secondary recycling, tertiary recycling might be considered the most cost-effective method of recycling materials while preserving environmental sustainability. In the tertiary recycling process, a solvent is used to recycle thermoplastic-based polymeric and textile materials, which may then be utilized in the thermal cracking, hemolysis, gasification, and thermal energy production processes. The solvent stimulates the polymerization of the thermoplastic chain, which essentially re-orients the form and chain structure of the thermoplastic matrix so that characteristics can be increased. It is possible to improve chemical and thermal sustainability by recycling procedures such as gasification and thermolysis.

Quaternary Recycling

After being tertiarily recycled, polymeric, and textile-based materials are put through a process called quaternary recycling. They are subjected to heat and burning to transform them into fuel and energy. Thermoplastic is a good alternative to energy-producing materials because of its high calorific value. The combustion process governs the quaternary recycling process for polymeric and textile-based products. Quaternary recycling involves the incineration of carpet waste to obtain thermal energy. While burning, this has higher ash content than that of coal. The problems with incineration or the generation of carpet-derived fuel (CDF) increase air pollution containing NOx emissions.

Table 1.2. Summary of recycling methods of carpet waste

Sr. No.	Types of carpet	Recycling method of Carpet waste	Ref.
1	Post-consumer carpet waste	VARTM	(Jian Wang and Mao 2012)
2	Carpet waste jute yarn	Compression molding technique	(Onal and Karaduman 2009)
3	Discarded nylon waste carpet	Compression molding technique	(Pan et al. 2016)
4	Nylon and Olefin waste carpet	VARTM	(Jain et al. 2012)
5	Carpet waste jute yarn	Compression molding technique	(Karaduman and Onal 2011)
6	Jute/Polypropylene	Compression molding technique	(Karaduman and Onal 2016)
7	Flax/jute	Compression molding technique	(Karthi et al. 2021)

Summary

Carpets are an integral part of almost all private and public spaces as they add coziness, warmth, and style to any space. Experts say that delayed methods and half-hearted approaches to dispose of garbage in general and carpet waste, in particular, are only delaying the problem rather than addressing it. A concerted effort, including the governments, major industries, and R&D funding, is needed to enable long-term planning on a global scale. This strategy could turn waste into valuable raw materials for industry, contributing to both sustainable growth and maintaining the natural order.

This chapter briefly describes the history of carpets on a global scale. It also discusses how the carpets are classified and what their uses are. The most prominent carpet belt regions throughout the world are also discussed in this chapter. As the chapter ends, a brief description of carpets and their critical functions in India are described too.

It is clear that a new method is required to develop carpet recycling to alleviate the problems generated through conventional techniques. Therefore, this book has made an effort to discuss the re-utilization of discarded carpets, which is quite different from the above recycling process. It focuses on the production of composite materials using waste carpets that can be used in low-structural applications.

References

Ahmed, Hemn Unis, Rabar H. Faraj, Nahla Hilal, Azad A. Mohammed, and Aryan Far H. Sherwani. 2021. "Use of Recycled Fibers in Concrete Composites: A Systematic Comprehensive Review." *Composites Part B: Engineering* 215: 108769. doi:10.1016/j.compositesb.2021.108769.

Bano, R., and Nida Fatima. 2014. "Role of Small Scale Industry and Women Empowerment: A Case Study in Carpet Industry of Bhadohi District, U.P." *International Journal of Managment, IT and Engineering* 4 (3): 152–64.

Binggeli, Corky. 2011. *Interior Graphic Standards*. Edited by Corky Binggeli. 2nd ed. John Wiley & Sons.

Chen, Yushu, Guotian Cai, Lixing Zheng, Yuntao Zhang, Xiaoling Qi, Shangjun Ke, Liping Gao, Ruxue Bai, and Gang Liu. 2021. "Modeling Waste Generation and End-of-Life Management of Wind Power Development in Guangdong, China until 2050." *Resources, Conservation and Recycling* 169. Elsevier B.V.: 105533. doi:10.1016/j.resconrec.2021.105533.

Cline, C. D., and J. D. Friddle. 1992. "Specialty Textile Coatings —a Formulation Overview." *Journal of Industrial Textiles* 22 (1): 32–41. doi:10.1177/15280 8379202200104.

Cunningham, Patrick R., Peter G. Green, and Sabbie A. Miller. 2021. "Utilization of Post-Consumer Carpet Calcium Carbonate (PC4) from Carpet Recycling as a Mineral Resource in Concrete." *Resources, Conservation and Recycling* 169: 105496. doi:10.1016/j.resconrec.2021.105496.

Dhawan, Ridham, Brij Mohan Singh Bisht, Rajeev Kumar, Saroj Kumari, and S. K. Dhawan. 2019. "Recycling of Plastic Waste into Tiles with Reduced Flammability and Improved Tensile Strength." *Process Safety and Environmental Protection* 124: 299–307. doi:10.1016/j.psep.2019.02.018.

Drescher, Daniela, Ingrid Zeise, Heike Traub, Peter Guttmann, Stephan Seifert, Tina Büchner, Norbert Jakubowski, Gerd Schneider, and Janina Kneipp. 2014. "In Situ Characterization of SiO2 Nanoparticle Biointeractions Using BrightSilica." *Advanced Functional Materials* 24 (24): 3765–75. doi:10.1002/adfm.201304126.

Ghobakhloo, Morteza, and Masood Fathi. 2021. "Industry 4.0 and Opportunities for Energy Sustainability." *Journal of Cleaner Production* 295: 126427. doi:10.1016/j.jclepro.2021.126427.

Goswami, K. K. 2009a. *Advances in Carpet Manufacture*. Edited by K.K. Goswami. *Advances in Carpet Manufacture*. 2nd ed. United Kingdom: Elsevier B.V. doi:10.1533/9781845695859.

Goswami, K. K. 2009b. *Developments in Handmade Carpets: Design and Manufacture*. *Advances in Carpet Manufacture*. Second Edi. Elsevier Ltd. doi:10.1016/B978-0-08-101131-7.00012-5.

Goswami, K. K. 2018. "Developments in Handmade Carpets: Introduction." In *Advances in Carpet Manufacture*, edited by K. K. Goswami, Second Edi, 213–68. United Kingdom: Elsevier Ltd. doi:10.1016/B978-0-08-101131-7.00011-3.

Henckens, Theo. 2021. "Scarce Mineral Resources: Extraction, Consumption and Limits of Sustainability." *Resources, Conservation and Recycling* 169: 105511. doi:10.1016/j.resconrec.2021.105511.

Islam, Shafiqul, and Gajanan Bhat. 2019. "Environmentally-Friendly Thermal and Acoustic Insulation Materials from Recycled Textiles." *Journal of Environmental Management* 251: 109536. doi:10.1016/j.jenvman.2019.109536.

Jain, Abhishek, Gajendra Pandey, Abhishek K. Singh, Vasudevan Rajagopalan, Ranji Vaidyanathan, and Raman P. Singh. 2012. "Fabrication of Structural Composites from Waste Carpet." *Advances in Polymer Technology* 31 (4): 380–89. doi:10.1002/adv.20261.

Jaiswal, Balram, Vijay Kumar Singh, Sanjay Mishra, and Rajesh Kumar Verma. 2021. "Study on Polymer (Epoxy) Composite Using Carpet Waste for Lightweight Structural Applications: A New Approach for Waste Management." *Materials Today: Proceedings* 44: 2678–84. doi:10.1016/j.matpr.2020.12.681.

Kalebek, Nazan Avcioglu. 2016. *Fiber Selection for the Production of Nonwovens*. Edited by Osman Babaarslan ED1 - Han-Yong Jeon. IntechOpen. doi:https://doi.org/10.5772/61977.

Karaduman, Y., and L. Onal. 2011. "Water Absorption Behavior of Carpet Waste Jute-Reinforced Polymer Composites." *Journal of Composite Materials* 45 (15): 1559–71. doi:10.1177/0021998310385021.

Karaduman, Y. and, and L. Onal. 2016. "Flexural Behavior of Commingled Jute/Polypropylene Nonwoven Fabric Reinforced Sandwich Composites." *Composites Part B: Engineering* 93: 12–25. doi:10.1016/j.compositesb.2016.02.055.

Karthi, N., K. Kumaresan, S. Sathish, L. Prabhu, S. Gokulkumar, D. Balaji, N. Vigneshkumar, S. Rohinth, S. Rafiq, S. Muniyaraj, S. Pavithran. 2021. "Effect of Weight Fraction on the Mechanical Properties of Flax and Jute Fibers Reinforced Epoxy Hybrid Composites." *Materials Today: Proceedings* 45 (9): 8006–10. doi:10.1016/j.matpr.2020.12.1060.

Kumar, Tejendra, Sanjay Mishra, and Rajesh Kumar Verma. 2019. "Fabrication and Tensile Behavior of Post-Consumer Carpet Waste Structural Composite." *Materials Today: Proceedings* 26: 2216–20. doi:10.1016/j.matpr.2020.02.481.

Liu, Fang, Hanqiao Liu, Na Yang, and Lei Wang. 2021. "Comparative Study of Municipal Solid Waste Incinerator Fly Ash Reutilization in China: Environmental and Economic Performances." *Resources, Conservation and Recycling* 169: 105541. doi:10.1016/j.resconrec.2021.105541.

Miraftab, M. 2018. "Recycling Carpet Materials." In *Advances in Carpet Manufacture*, edited by K. K. Goswami, Second Edi, 65–77. United Kingdom: Elsevier Ltd. doi:10.1016/B978-0-08-101131-7.00005-8.

Mirzababaei, M., M. Miraftab, M. Mohamed, and P. Mcmahon. 2013. "Impact of Carpet Waste Fibre Addition on Swelling Properties of Compacted Clays." *Geotech Geol Eng* 31: 173–82. doi:10.1007/s10706-012-9578-2.

Mishra, Kunal, Sarat Das, Ranji Vaidyanathan, and Tooling Materials. 2019. "The Use of Recycled Carpet in Low-Cost Composite Tooling Materials." *Recycling* 4: 12 (1-8). doi:10.3390/recycling4010012.

Mohammadhosseini, Hossein, A. S. M. Abdul Awal, and Jamaludin B. Mohd. 2017. "The Impact Resistance and Mechanical Properties of Concrete Reinforced with Waste Polypropylene Carpet Fibres." *Construction and Building Materials* 143: 147–57. doi:10.1016/j.conbuildmat.2017.03.109.

Mohammadhosseini, Hossein, Mahmood Tahir, Abdul Rahman, Mohd Sam, Nor Hasanah, Abdul Shukor, and Mostafa Samadi. 2018. "Enhanced Performance for Aggressive Environments of Green Concrete Composites Reinforced with Waste Carpet Fibers and Palm Oil Fuel Ash." *Journal of Cleaner Production* 185 (1): 252–65. doi:10.1016/j.jclepro.2018.03.051.

Mohapatra, Himansu Shekhar, and Rajesh Kumar Verma. 2017. "Indian Handmade Carpet- A Millennium Floor Covering." *Journal of Textile Association* 3: 157–62.

Onal, L., and Y. Karaduman. 2009. "Mechanical Characterization of Carpet Waste Natural Fiber-Reinforced Polymer Composites." *Journal of Composite Materials* 43 (16): 1751–68. doi: 10.1177/0021998309339635.

Pan, Gangwei, Yi Zhao, Helan Xu, Xiuliang Hou, and Yiqi Yang. 2016. "Compression Molded Composites from Discarded Nylon 6/Nylon 6,6 Carpets for Sustainable Industries." *Journal of Cleaner Production* 117: 212–20. doi:10.1016/j.jclepro.2016.01.030.

Realff, Matthew J., Jane C. Ammons, and David Newton. 1999. "Carpet Recycling: Determining the Reverse Production System Design." *Polymer-Plastics Technology and Engineering* 38 (3): 547–67. doi:10.1080/03602559909351599.

Sahana, S. 2021. "Waste Management in India." *Shanlax International Journal of Arts, Science and Humanities* 8: 283–87. doi:10.34293/sijash.v8is1-feb.3968.

Siddique, Rafat, Jamal Khatib, and Inderpreet Kaur. 2008. "Use of Recycled Plastic in Concrete: A Review." *Waste Management* 28 (10): 1835–52. doi:10.1016/j.wasman.2007.09.011.

Sotayo, Adeayo, Sarah Green, and Geoffrey Turvey. 2015. "Carpet Recycling: A Review of Recycled Carpets for Structural Composites." *Environmental Technology and Innovation* 3: 97–107. doi:10.1016/j.eti.2015.02.004.

Teli, M. D. 2018. "Finishing of Carpets for Value Addition." In *Advances in Carpet Manufacture*, edited by K. K. Goswami, Second Edi, 175–211. United Kingdom: Elsevier Ltd. doi:10.1016/B978-0-08-101131-7.00010-1.

Verma, R.K., P. Dwivedi, and S. Sinha. 2017. "Simultaneous Optimization in Quality Characteristics of Machine-Made Tufted Carpets." *Melliand International* 23 (3): 159–61.

Wang, Jian, and Qianchao Mao. 2012. "A Novel Process Control Methodology Based on the PVT Behavior of Polymer for Injection Molding." *Advances in Polymer Technology* 32 (S1): E474–85. doi:10.1002/adv.21294.

Wang, Junxiong, Jiakuan Yang, Huijie Hou, Wei Li, Jingping Hu, Mingyang Li, Wenhao Yu, Zhongyi Wang, Sha Liang, Keke Xiao, Bingchuan Liu, Kai Xi, R. Vasant Kumar. 2021. "A Green Strategy to Synthesize Two-Dimensional Lead Halide Perovskite via Direct Recovery of Spent Lead-Acid Battery." *Resources, Conservation and Recycling* 169: 105463. doi:10.1016/j.resconrec.2021.105463.

Wang, Youjiang. 2007. "Carpet Fiber Recycling Technologies." In *Ecotextiles: The Way Forward for Sustainable Development in Textiles*, edited by M. Miraftab and A. R.

Horrocks, 1st ed., 26–32. Cambridge England: Woodhead Publishing Limited. doi:10.1533/9781845693039.1.26.

Wang, Youjiang, Mehmet Ucar, and Youjiang Wang. 2014. "Utilization of Recycled Post Consumer Carpet Waste Fibers as Reinforcement in Lightweight Cementitious Composites." *International Journal of Clothing Science and Technology* 23 (4): 24–248. doi:10.1108/09556221111136502.

Wilson, David C. 2007. "Development Drivers for Waste Management." *Waste Management & Research* 25 (3): 198–207. doi:10.1177/0734242X07079149.

Xanthos, M., S. K. Dey, S. Mitra, U. Yilmazer, and C. Feng. 2002. "Prototypes for Building Applications Based on Thermoplastic Composites Containing Mixed Waste Plastics." *Polymer Composites* 23 (2): 153–63. doi:10.1002/pc.10421.

Chapter 2

Novel Methodology in the Fabrication of Composite from Carpet Waste

Polymeric materials have been extensively used in the industrial sector for the past three decades to meet changing consumer and market demands. Due to their remarkable qualities, including reduced weight, corrosion resistance, and improved mechanical performance, polymeric composites successfully replace traditional production materials today. Utilizing composites composed of polymers and polymers modified by fillers has a number of advantages. In contrast to traditional metallic materials, it has distinctive properties such as low density, improved strength, simplicity in production, and low cost. The mechanical properties of the existing polymer composites still need to be improved. The mechanical properties of the existing polymer composites still need to be improved. Different types of fiber were added to these polymer-based composite materials to strengthen them(reinforcement materials). Reusing waste materials to create affordable and ecologically friendly production materials is another issue for the manufacturing sector. According to environmental regulations and conditions, the non-biodegradable waste produced by industries becomes a top problem. The non-biodegradable waste generated from the industries becomes a prime concern for decomposition according to environmental rules and conditions. The fibrous debris produced by the carpet industry is a significant source of solid waste that must be dumped in a broad region, which is an expensive process. In this regard, researchers seek to create modified polymer composites from carpet waste. It has been noted that it is in the progressing stages and requires increased interest from academic institutions, research organizations, manufacturing companies, and government agencies.

Solid waste is a major problem for society that affects the growth of various pollution causes and environmental worries. Reusing heavy waste lots has become a major problem for the manufacturing industry despite numerous approaches to solid waste management being studied by eminent experts. The strength and stiffness of polymer composites are quite good when combined with low-density components. The high-density polymer materials are appropriate for weight-saving demands on a variety of items. The composite still consists of various macro-level materials, so it cannot be considered traditional engineering material with a homogeneous structure. Composites cannot be regarded as conventional engineering materials with homogeneous structures because they nevertheless contain a variety of macro-level constituents.

Composites

The term composite can be defined as "material systems" that are multifunctional and have properties that differ from discrete phases, *viz.*, matrix phase and reinforced phase. It can also be defined as the combination of two or more materials to form a cohesive structure with improved properties. In other words, a composite is a combination of two or more elements, one of which is rigid material reinforcement, and the other is a binder or matrix that keeps the reinforcement in place (shape/size). When these phases (reinforcement and matrix) are combined, the final composite material retains its individual identities (both reinforcement and matrix), substantially affecting the composite performance. The term composite refers to a material system made up of two or more materials with improved material properties than the individual material (Mohammed et al. 2015). The structure of constituents is combined so that the components retain the form of a chemical compound in their respective functional phases (Horoschenkoff and Christner 2012). The exceptional engineering properties of polymers attracted academia and industry professionals to explore the composites research aspect in recent decades. The composite material phase begins with a low-density element and progresses to a high-strength element. Composite materials are primarily composed of two phases: matrix and reinforcement. The matrix is a primary phase with a continuous form. It has a more ductile nature and a softer phase. The role of the matrix phase is to keep the secondary phase (reinforcement) in the desired shape and size under loading conditions. These phases have a discontinuous form entrenched in the matrix in specific ratios to enhance the physical and mechanical properties. These are less ductile and stronger than the matrix. It is also called as reinforcing phase.

Classification of Composites

In this section, a broad classification level and their significance have been described in a detailed manner. Generally, composite materials are classified into two subsections: composite based on matrix and composite based on reinforcement.

Classification Based on Matrix Materials

The matrix can usually categorize composite materials into three sections. These are composite in Metal Matrix (MM), Ceramic Matrix (CM), and Polymer Matrix (PM) (Gay, Hoa, and Tsai 2002). Composites in the metal matrix have several benefits over monolithic metals, such as a higher module, higher specific strength, improved temperature properties, and less thermal stability. Because of its high-strength properties, MM composite is being considered for a wide range of applications, including compressor blades, nozzles, steam turbines, and load-bearing (Basu and Debnath 2015; Singh 2014). One of the primary goals of manufacturing ceramic matrix composites is to improve durability. Usually, ceramic matrix composites can increase strength and stiffness simultaneously (Koch et al. 2006; Jannatyha et al. 2020).

Because of their enhanced properties like high strength-to-load ratio, anti-corrosive properties, dimensional tolerances, high elasticity, and damping coefficient, hybrid polymeric materials have become an essential part of modern production systems that are in use today. A new generation of composites is produced by synthesizing innovative lightweight composites with high mechanical characteristics. In this series, polymer (epoxy) based composites are extensively used due to their advanced level of cross-linking and high aspect ratio. Over the last few decades, manufacturing sectors have seen that customers are satisfied with various polymer composite materials. Polymers play a significant role in composite families because of their favorable properties. Polymer composites are widely used in producing aerospace, marine, space, and automotive components due to their enhanced mechanical properties and modified fatigue life (Rubino et al. 2020). It becomes a suitable replacement for conventional engineering materials like non-metallic and metallic alloys. The physical and structural characteristics of the materials used in production, such as low point-specific weight and strong resistance to deterioration, must be distinctive (Brebu 2020; Uygunoglu, Gunes, and Brostow 2015). In these multifunctional situations, it can ensure both safety and economic performance. In composites, the matrix material is the primary phase that surrounds and offers its bulky shape to the polymer. The polymer matrix phase is critical for load-bearing and load transfer between its cross-link connections. Therefore, their products' low weight and versatility make them significant, therefore, considered a highly desired element for the aviation industry or applications with low specific weight components. Polymers can be separated into thermoplastic composites and thermoset composites based on the properties of the polymer matrix. Thermoplastics are temperature sensitive since they can liquidate at temperatures above their melting point during the composite manufacturing process. The individual molecules in thermoplastics are not bonded chemically.

On the other hand, Van der Waals secondary bonds are formed by weak intermolecular forces (Mohammed et al. 2015). Lack of a hard binding force and adequate heat energy can be used to interrupt the thermoplastic materials. With the addition of heat, the molecules' tendency to move relative to each other allows for further transition and reconfiguring. The material solidifies when the heat source is removed; therefore, it may be reconfigured or recycled with the required application of pressure and heat. Thermoplastic products seem more ductile than other composites because of these distinguishing features. There is a realistic limit to the amount of thermoplastic reprocessing due to rising oxidation in the molten state. Subsequently, the liquid resin has solidified, and thermoset materials undergo a chemical reaction during manufacturing. In this condition, cross-linking chemically bonded to the molecules together, resulting in a rigid, three-dimensional network framework and an irrevocable composite. Smaller unlinked molecules known as monomers are found in the resin material, which is made of carbon. The chemical reaction is accelerated by applying a curing agent, also known as a hardener, which is nothing more than a catalyst. Oxidized polymer, polyvinyl, polyurethane, and phenolic are widely used in thermoset matrices.

These are composed of matrices derived from polymer-based materials like thermosets or thermoplastics (Q. H. Qin 2015). The materials which come under thermoset are unsaturated polyester, epoxy, etc., which come under thermoplastic are polycarbonate, polyvinylchloride, nylon, polystyrene, etc. Others that could be considered include PMCs, embedded glass, carbon, etc. PMC consists of fibers as reinforcing material which is comparatively much stiffer and more robust than the matrix. It also becomes attractive due to its lightweight properties, which can be modified for any particular application. A metal matrix composite is a composite material made up of at least two constituent elements, one of which must be metal and another metal, ceramic, or organic ingredient. A hybrid composite consists of three or more materials. These comprise a ceramic-based matrix with embedded fibers of any ceramic materials or other material in the dispersed phase. Reinforcement in CMC can be short (discontinuous fiber) or long fiber (continuous fiber). Silicon carbide (SiC) is mainly preferred as a reinforcing material because of its high strength and stiffness. It is used in high-temperature applications but has poor tension and shear properties. Figure 2.1 illustrates the classification of composite material based on the matrix phase.

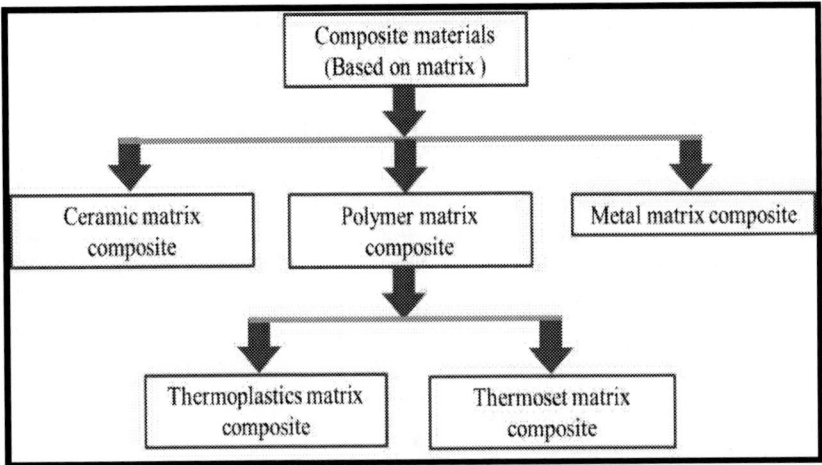

Figure 2.1. Types of composite materials based on matrix (Vijay, Rajkumara, and Bhattacharjee 2016).

Classification Based on Reinforced Material

The structure constituent determines the intrinsic configuration of the composite. It links the dispersion and matrix phases in the form of bonding. Based on the reinforcement, composite can be categorized into three groups. (a) Particulate Composites, (b) Fiber Reinforced Composite, and (c) Structural Composite. The composite particles are of different forms and dimensions and are randomly distributed in the matrix. Short-fiber composites are the discontinuous type used in composite reinforcement materials. The diameter of the fiber here is very small in contrast with the length of the reinforcement. The fibers can be distributed uniformly to create a one-way structure.

Continuous fibers are incorporated into the composite and bear uniaxial strength, resulting in increased material strength.

The name suggests that the reinforcement in such a composite is a particle in nature. The particle in this composite can be cubic, spherical, tetragonal, or any other regular or irregular shape, but the condition is approximately equiaxed. These particles help enhance the properties of composites, such as creep, wear, abrasion resistance, etc. Fibrous materials are distinguished by having lengths significantly greater than their cross-sectional area. However, the dimension of the reinforcement determines its contribution to the composite's properties. Fibers significantly improve the matrix's fracture resistance because their long size discourages crack growth. The surface area of the fiber matrix is kept as large as possible for better load transfer to the fiber composite. Structural composites are layered composites that are generally low-density composites and used in applications requiring structural integrity, high tensile strength, high stiffness, etc. These composites are comparatively strong and lightweight. The properties of composites are dependent on constituents and the geometrical design of the structural element. Figure 2.2 depicts the reinforcement-based classification of composite materials.

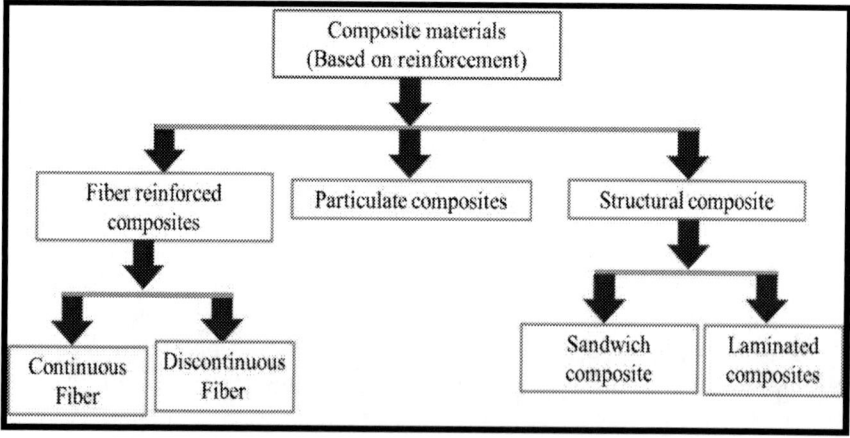

Figure 2.2. Classification of composite materials based on reinforcement(Vijay, Rajkumara, and Bhattacharjee 2016).

Recycling Process

Reusing waste offers a practical means of creating affordable products and lowering environmental risks. There aren't many studies on carpet recycling for waste management and composite creation. Today, it is challenging to recycle a variety of carpet trash found in landfills. Ingenious attempts have been performed to make use of this garbage in a variety of ways. Ingenious attempts have been performed to make use of this garbage in a variety of ways. The problems associated with this waste can be resolved by increasing its proper utilization. The management of such wastes has been the subject of ongoing studies that might offer an alternative approach. Eminent academics carried out some preliminary studies into the recycling of solid waste produced by manufacturing industries.

In this series, Bateman et al. (Bateman and Wu 2001) used carpet waste to convert it into an efficient product. The author was concerned about waste carpets and low-density polyethylene in this study (LDPE). Maleic anhydride (MAh) PE copolymer as a compatibilizer was taken in small amounts. This study used carpet fibers to reinforce polythene and concluded that it could be composite. Atakan et al. (Atakan, Sezer, and Karakas 2018). The author looked at a recycled polyethylene terephthalate (rPET) manufactured from a PET bottle and found that the mechanical and fastness properties of molded carpets made with rPET were superior. Haines et al. (Haines et al. 2020) utilized shredded carpets from demolished houses as fuel in cement kilns, which have a high heat value similar to low-grade coal. They discovered that the quantity of carbon monoxide and other non-burned substances did not alter much. They discovered that nylon fibers generated a significant increase in nitrogen oxide (NO). Rushforth et al. (Rushforth et al. 2005) utilized recycled granulated carpet tiles to investigate the soundproof performance material. Sifting granulated waste has increased acoustic and sound insulation properties. Karaduman et al. (Onal and Karaduman 2009) examined the study of dry and water absorption carpet waste composite. They found that the wet composite's flexural strength and modulus decreased by 40% and 65%, respectively, but its impact strength increased by 30%. Kocak et al. (Kocak, Akalin, and Merdan 2016) developed a novel hybrid composite by combining carpet waste with cement. They discovered that while their impact strength stayed constant, their flexural strength increased with density. Sotayo et al. (Sotayo, Green, and Turvey 2018) conducted experiments and converted wool waste face fiber into a novel composite. Composite was employed as an alternative to wood and PVC for fence posts and rails. The innovative composite's average flexural strength was twice as strong as its tensile strength. Alrshoudi et al. (Alrshoudi et al. 2020) developed the composite from waste carpet and utilized it in construction. They discovered that the combination of palm oil fuel ash and polypropylene carpet material produced a highly profitable product. It shows how the fiber and matrix collaborated to increase the material's flexural and tensile strengths as well as its elasticity, energy absorption, and impact resistance. Dayiary et al. (Dayiary, Najar, and Shamsi 2009) conducted a study based on energy storage under constant compressive stress. The theoretical model was used to perform the compression test on the cut-pile carpet using a variety of fibers, including PP, acrylic spun, and other fibers. It was found that the proposed theoretical model has a good relationship with the experimental results. Awal et al. (Awal and Mohammadhosseini 2016) processed a novel composite by mixing the polypropylene waste carpet fiber and plain concrete. It was determined that adding carpet fiber to concrete significantly decreased the concrete's workability and density while concurrently increasing its flexural and splitting tensile strengths. Miraftab et al. (Mohsen Miraftab, Horrocks, and Woods 1999) developed a hybrid composite by incorporating waste nylon carpet fiber into the clayey soil. In contrast to the conventional clayey or inferior soils, this combination demonstrated greater load-bearing capacity, internal cohesion, shear strength, and comprehensive strength, according to the author's results. The increase of the aforementioned qualities largely depends on the soil's moisture content. The kind of fiber may influence the results of such a constructed composite, the length of the fiber, the type of soil, and the environmental conditions in order to increase the attributes of glass fiber reinforced polymer (GFRP). Wang (Y. Wang 2006) studied improved FRC properties by incorporating recycled nylon fibers from industrial carpet waste. The results revealed improvement in shatter resistance, ductility, and energy absorption compared with virgin polypropylene (PP) fibers used earlier. Gowayed et al. (Gowayed and El-halwagi 1995) prepared the composite material using PP carpet waste fabric as reinforcement and polyethylene (PE) as matrix material. They reported that fabricated composite revealed enhanced mechanical and chemical properties than that of pure polyethylene. Jain et al. (Jain et al. 2012) executed structural composites from the waste carpet through the resin transfer molding (RTM) technique in a vacuum environment. Two types of carpets, such as nylon and olefin-based, were used for fabrication purposes. This paper included characterization such as flexural test and SEM of composites and their applications, including a low-cost container, civil infrastructure components, and materials for impact protection. This technique obtained an improved mechanical property of carpet composites compared with other recycled carpet-based materials. Islam et al. (Islam and Bhat 2019) conducted experiments and produced insulation composites from recycled textiles for acoustic and thermal uses. This played an important role in energy reserves and the decrease of environmental pollution. One sector which produces a large amount of waste is the textile sector, though these wastes contain valuable fiber products. These wastes can be recycled to make various products for several applications, including the thermal and acoustic applications mentioned above. This recycling process helped in the reduction of waste and the fight against severe environmental issues. Xanthos et al. (Alrshoudi et al. 2020) fabricated a low-cost thermoplastic composite with consistent properties by taking nylon-6 fibers from used carpets and thereby used them for building applications. It can be used as a replacement for wood thermal barriers used in steel-based stud assembly. Flexural strength, modulus, and strength of LDPE decreased by the presence of automotive shredder residue (ASR). Strength and strain at break decreased, but the modulus of LDPE in tensile and flexural loading increased with the addition of carpet residue. Carpet residue-based composite shows better results than ASR-based thermal barrier in terms of higher density, reduced creep rate, thermal conductivity, leachates, and flammable properties. Zareei et al. (Alireza et al. 2019) investigated the change in tensile, compressive as well as flexural strengths when recycled waste ceramic aggregate (RWCA) and waste carpet fiber (WCF) were combinedly utilized in high-strength concrete (HSC). It was found that when RWCA replaced 40% natural coarse aggregate (NCA), it increased the compressive, tensile, and flexural strength by 13%, 15%, and 3%, respectively. WCF was added to the same mixture. It increased the tensile strength up to 21%. Combinedly showed an increase in water absorption by 48%. Miraftab and Mirzababaei (M Miraftab and Mirzababaei 2009) summarized almost all possible theories, which could be provided as a bunch of information collected. The area of topics that he covered started with the introduction, which included generation and concern for carpet waste. This paper had different pieces of information on the categories of carpets in which post-consumer carpets were focused more. Now, the challenging work related to recycling such waste carpets was explained. Future scope, sources, government role, and consumers regarding carpet waste as a matter of concern were all discussed. Thakur and Thakur (Thakur and Thakur 2014) fabricated polymer composites based on cellulose fibers of different varieties. The use of cellulose fibers as reinforcement components in other polymers helps develop specific properties into the product.

The reason behind this being an attraction for the researcher is its ease of availability, friendliness to the environment, and ease of processing applications in the field from automotive to medical. Kumar et al. (T. Kumar, Mishra, and Verma 2019) developed a structural application composite from the waste carpet using the VARTM fabrication method.

This attempt was performed to convert such waste into a mechanical application form. Using such composites as structural elements in the context of this study results in a noise barrier that assumes the form of a noise absorption coefficient. For several frequencies, it was found that the noise absorption of fiber composite was better than traditional, and this improvement applied to some frequency ranges. Gowayed et al. (Gowayed and El-halwagi 1995) fabricated composite using waste fabrics and plastics. The composite synthesized was an environmentally benign fiber composite material. Polypropylene "PP" was chosen as the reinforcing material, and polyethylene "PE" was selected as the matrix material. These were chosen due to their abundant availability. It was found that the average tensile modulus initially for pure ethylene was 0.15 GPa and its average tensile strength was 9 MPa. Chemical treatment of the fabrics of "PP" was done for the promotion of "wetting" characteristics of "PP" to enhance the surface compatibility between PP fabrics and PE resin. Fabrication of composite panels was done by compression molding machine and tested for different mechanical properties.

A variety of applications can benefit from using recovered fabrics. The amount of garbage disposed of, which is the component of the textile industry that contributes the most to carbon emissions, may be decreased using this technique. The detrimental consequences of landfills and burning on the local environment are avoidable. Previous studies indicate that recovered carpets may be chemically handled in many ways. The acidic sites on the carbon nanofiller are what make recycling more economical and accessible rather than a recycling process, which would be a significant step toward cost reductions. Given the preceding, the objective of the current study is to enhance the qualities of a composite made using waste carpet and polymer. The research on this issue is extremely important for recycling used carpet fibers into materials that are both functional and structural components. Researchers have been interested in using carpets and other waste materials that may be used for any type of good purpose. They have attempted to use scraps since they are easily accessible, and their goal has been to employ a variety of ways to modify them in a useful way. The blending of carpet fibers with other materials, the creation of composites, and the characterization of their properties all contribute to the efficient use of trash. The manufacturing and advanced mechanical investigations of the polymer composite made from carpet waste have received relatively little attention, according to the research that has been done so far on the subject of carpet waste management. The results of previous studies have primarily focused on the macro dimensions reinforcing material in the matrix phase. Industries like carpet and textile production must reuse or recycle the trash they produce in order to benefit society and the economy. From the perspective of waste management, this is significant. These industries generate a massive amount of garbage that is very heavy in weight. Due to the material's high cost, non-biodegradable nature, and environmental issues, the carpet business cannot use the standard recycling process. Because of this, the focus of this research was mostly on recycling used carpets to make polymer composites. An effort has been made to reduce the quantity of garbage generated as a result of inputting new carpets by repurposing old carpets. The development of lightweight structural composites for use in toys, frames, wall tiles, sound barriers, and other applications could be a successful waste management tactic that aids in achieving the sustainability aim.

Composite Development from Carpet Waste

Polymer-based composite materials weren't strong enough to handle the job in many engineering applications. In this framework, the physio-mechanical performance of polymers was effectively enhanced by the supplement of natural and synthetic fibrous reinforcement (Pathak et al. 2019; Shagor, Abedin, and Asmatulu 2021). The type of additives can augment the required manufacturing performance of the FRP product in the binding matrix. Fibers are the most common additive to improve bond strength, mechanical and electrical properties, etc. (Kunrath et al. 2019; Hosur, Mahdi, and Jeelani 2018). Polymer composites have been produced to use waste and discarded materials in some real-world multifunctional components (Siwal et al. 2020; M. Kumar, Saini, and Bhunia 2020). The fabrication of high-performance composite materials should have higher mechanical, acoustics, and physical properties (water absorption and fire resistance). Fiber has high tensile strength, longer and more proximal, with adequate interfacial bonding (Shin et al. 2017).

Thus, the longer and highly oriented fiber must be recorded to have high engineering properties (Gangineni et al. 2019). The shear strength of the fiber only depends on the average strength and width of the epoxy bonding interface. Also, the degree of randomness of the fiber in the matrix and the bonding strength affect overall performance (Naebe et al. 2014). Thus, it looks for specific tensile, shear, and bending values. It is about the benefits of carpet fiber reinforcement composite; the mechanical strength is higher than the plain materials. Thus, these high potential properties make the waste fiber-based polymer composite distinctive from the other materials. It is considered a suitable candidate for multifunctional component development in the manufacturing industry. In addition to thermoset polymers, cross-linked assembly pays for higher stiffness and physio-mechanical performances. The cross-linked nature, anisotropic, and non-homogeneity of fiber composite enhance the required engineering properties. In the epoxy resin, fiber materials reinforcing was observed as an active procedure to increase the strength, durability, and stiffness under different loading conditions. The pristine composite existing performances could be significantly enhanced through the modification of macro reinforced (fiber). Sometimes, recycling theory is used in the management of waste materials to investigate the pollution aspects. Limited investigational work has been conducted on recycling waste materials into structural composite applications. It necessitates more consideration for the effective application of carpet waste for the development of polymers in multifunctional products.

Composite laminates are becoming increasingly used in a wide variety of technological disciplines due to their unique mechanical and physical features. Composite laminates are particularly advantageous in the automobile industry, in addition to the production of sports apparatus, aircraft, aerospace products, marine ship goods, and petroleum products. Composite laminates also have several other applications. (Mohammed et al. 2015; R.-M. Wang, Zheng, and Zheng 2011). The leading method in the study of recycling carpet fiber composite laminates emphasizes using components with different assets (Nijssen and R.P.L.Nijssen 2015). The novel growth in structural composites commands a weight saving of around 50% by replacing the main structural elements with discarded carpet composites (A. Kumar, Sharma, and Dixit 2019; Myagkov et al. 2014). Utilizing a lightweight and high-strength composite is crucial if future developments in engineering technologies reduce waste and consumption while simultaneously improving acoustic performance. (Stewart 2009; Nonobe 2017). For required applications, polymer composites have become a direct replacement because of exceptional mechanical, chemical, and other performances, including low weight, stiffness to weight, fatigue, and thermal coefficient compared with metals (Stewart 2009; Myagkov et al. 2014; Nguyen, Nguyen, and Bach 2019).

Moreover, by sensibly varying fiber weight percentage, the fiber alignment of individual yarn in laminates affects the mechanical performance. Also, accordingly composite material properties can be altered to several ensemble requirements (Liu, Tang, and Cong 2012). Based on the particular composites and manufacturing procedures, the carpet includes wool, polypropylene, polyester, etc. (Mohammed et al. 2015; George, Sreekala, and Thomas 2001). The most popular material used for reinforcement is polyester (waste materials). A significant fraction of the laminates are made of structural composite materials. (Papageorgiou et al. 2019). Manufacturing industries and leading experts are very interested in polymer matrix composites because of their increased durability and simple manufacturing process compared to other conventional products. A typical matrix component found in composite materials is epoxy (resin). It benefits from non-volatility, better thermal and spatial flexibility, and strong bonding. (Zhao et al. 2016).

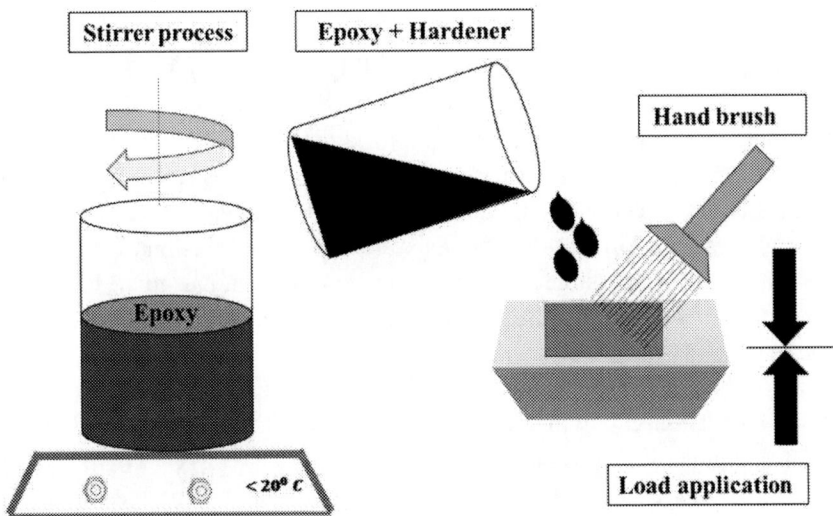

Figure 2.3. Hand layup process (J. Kumar, Verma, and Mondal 2020).

The composites were fabricated using an epoxy matrix. Due to the high viscosity of epoxy, it is challenging to make a spread over fiber reinforced (J. Kumar et al. 2021; Kesarwani et al. 2021; J. Kumar, Verma, and Mondal 2020). Composite material is made up using epoxy as a primary and secondary phase waste carpet fiber. The epoxy resin and fiber transfer the strength &

mechanical features. Resin (density:1.162 gm/cm^3) was stirred for up to 30 minutes at 60 °C. The hardener-D with an equivalent ratio of 70:30 was added after 30 minutes, and the mixture was stirred for 1 hour at 27 °C. Finally, the mixture was laid up onto the discarded fiber layer by layer into a square shape after the desired thickness load was assigned using the method of hot pressing (3 MPa,170°C, 200 min). The development of discarded composite samples procedure for fabrication is mentioned in Figure 2.3. The following steps were used for the fabrications phase:

- The desired amount of epoxy resin was collected and heated on a hot stirrer plate at about 75°C to decrease its viscosity.
- The prepared resin and hardener were mixed and again stirred for about 1 hour.
- The Conventional (Hand layup) technique was used for the fabrication process (Dubey et al. 2021; J. Kumar, Abhishek, and Xu 2022; J. Kumar and Verma 2021).

Vacuum-Assisted Resin Transfer Molding (VARTM) Method

To boost productivity and fabrication of high-quality goods, the precise manufacture of defect-free structural composites is necessary. This research focuses on the creation of a modified composite using vacuum-assisted resin transfer molding (VARTM) technology. The efficient infusion of resins into the matrix is made possible by this vacuum environment. Due to the combined properties of both materials, the product can be endorsed for a structural component end use such as sound insulation, noise barrier, wall tiles, and so forth. Eminent scholars proposed the VARTM fabrication process to develop waste carpet-based polymer composites (J. Kumar, Kumar, Jaiswal, et al. 2022; J. Kumar, Kumar, Kumar, et al. 2022).

This study relates to the re-utilization of waste carpets for developing the lightweight structural polymer composite. Generally, the life of any carpet is typically 10 to 15 years, after which it is discarded as carpet waste (Sohn 1997; Street et al. 2008). There are generally two types of carpet waste, post-consumer carpet waste, which includes carpets from end-of-life waste streams. Pre-consumer carpet waste includes scraps or offcuts generated during their manufacture and installation. There are numerous types of carpet materials, such as nylon, wool, polyester, and polypropylene. Carpet wastes are generated by households, hotels, cars, buses, furniture, movie theatres, etc.

These various carpet waste materials are used as reinforcement and epoxy resin as a matrix for the structural composite material. Epoxy improves the interfacial adhesive characteristics between the reinforcement and matrix of the composite's material. The composites developed through the VARTM technique ensure the void-less infusion of the matrix in the carpet. This discloses the productive and cost-effective approach for using carpet waste in the composite material.

Additionally, this method refers mainly to the management of solid carpet industry waste and post-consumer carpet waste. This chapter explains the precise fabrication procedure of composite material from waste carpets. The flow chart of methods and processes is elaborated in Figure 2.4.

Figure 2.4. Flow chart for the proposed methodology.

For the development of composite, the waste carpet has been collected from different sources of pre-consumer scraps industry and post-consumer like households and automobiles. The epoxy material is a thermoset resin used in polymer composites for load distribution to reinforcement applications. The epoxy has significant advantages like fabrication at room temperature, excellent composite performance at elevated temperature, superior interlaminar shear strength properties, low shrinkage, resistance to a corrosive environment, and robust mechanical properties. For the fabrication of composite samples, epoxy was used as a polymer phase. A solvent-free, low-viscosity, bisphenol-A-based epoxy resin (Araldite GY-257) and hardener (Araldite HY-951) were used for matrix materials. The properties of epoxy resin have dissipated in Table 2.1.

Table 2.1. Specification of Epoxy resin (Araldite GY-257) (Kumar Verma et al. 2021)

Parameters	Value
Density	1.15gm/cc
Viscosity	450-650 mPa.s at 25 °C
Curing	25 °C 20-30 hrs.
Specific gravity	1.8

Certain items and tools are required for the fabrication setup, such as a vacuum bag, mesh, peel ply, spiral tube, clamp, resin tube, sealant tape, vacuum pump, etc. The release agent, silica gel spray, was utilized to remove the developed specimen simply from the bottom surface. These carpet piles provide mechanical strength, whereas epoxy resin acts as a binding material and distributes the reinforced composite load. Polymer composite fabrication was done with the waste carpet (300 x 250 mm^2) in FBBF and BFFB configuration by the infusion of thermosetting epoxy resin.

The VARTM technique has been used for polymer composite fabrication, which has the advantage of uniform resin flow and minimum porosity while preparing the sample. The fabrication setup and block diagram are shown in Figures 2.5 and Figure 2.6. Initially, the acquired waste carpets are gently washed, removing all the dust particles. Cleaned carpets are dried in the open sunlight for 24 hrs. After the dying process, the carpet is cut in the dimension of 300×250 mm^2. The epoxy resin is used as a polymer matrix for the composite development of epoxy diglycidyl Bisphenol-A (Araldite GY 257), low-viscosity Tri-ethylene-tetra-amine (Araldite HY 951) used as a curing agent. The ratio of epoxy and the hardener is 10:1 ratio by weight. The carpet pile was laid down in a vacuum bag into both configurations, BFFB and FBBF, separately, and an airtight mold was prepared with the help of tacky tape, polyethylene sheet, inlet, and outlet pipe. The vacuum pressure is applied with the vacuum pump of -0.85 atm pressure for the proper infusion of resin in the carpet pile through the spiral wrap. After appropriate infusion from the carpet pile, both the inlet and outlet pipe are clamped tightly and left for 30 hours for proper curing. Table 2.2 provides an overview of the progress made in developing carpet waste polymer composite materials. All the processes are conducted at a room temperature of (approx. 25 °C).

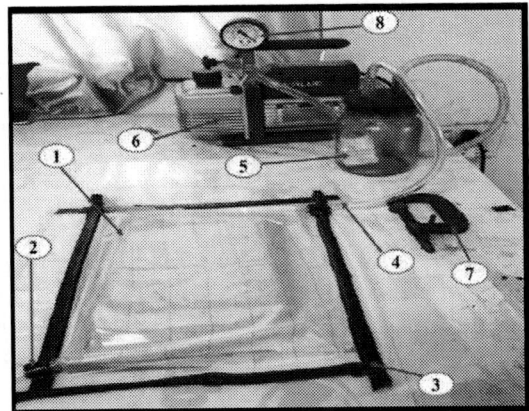

1. Vacuum bag
2. Inlet valve
3. Sealant tape
4. Outlet valve
5. Vacuum box
6. Vacuum pump
7. Clamp
8. Air pressure gauge

Figure 2.5. Fabrication setup for the development of waste carpet composite (Kumar Verma et al. 2021).

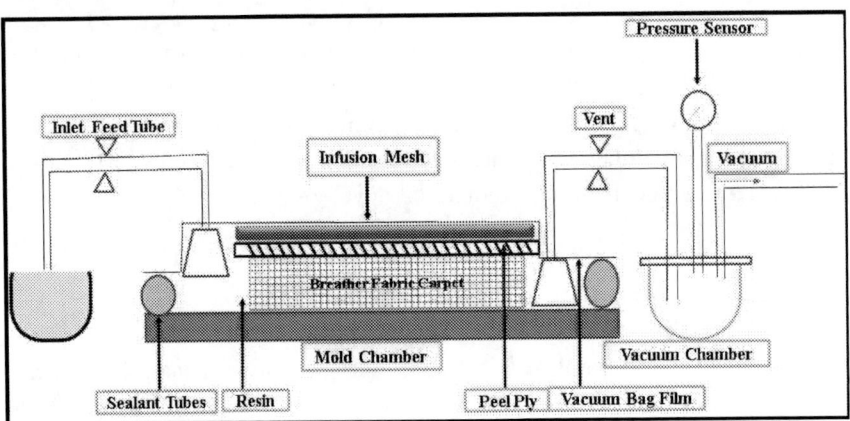

Figure 2.6. Block diagram of VARTM setup (Bhatt, Gohil, and Chaudhary 2022).

Table 2.2. Summary on composite development from carpet waste using VARTM technique

Sr. No.	Reinforcement	Matrix	References
1	Post-consumer carpet/Graphene	Epoxy	(Mishra et al. 2019)
2	Nylon/Nano clay	Epoxy	(Das 2009)
3	Post-consumer carpet	Epoxy	(T. Kumar, Mishra, and Verma 2019)
4	Polyester	Epoxy	(Kedari, Farah, and Hsiao 2011)
5	Waste carpet	Epoxy	(J. Kumar, Kumar, Jaiswal, et al. 2022)

Summary

This chapter gives a brief overview of recycling strategies used to reduce the production of solid waste and environmental contamination. It provides a succinct explanation of composite materials, their classification, and comparable crucial engineering roles, followed by an overview of carpet waste utilization in engineering and the significance and challenges for the development of polymer composites. This chapter also discusses the importance of recycling techniques for the progress of discarded polymer composite materials and includes a literature survey of current and previous work related to fabrication.

This chapter discusses sample preparation methods (hand layup) and the processing of carpet wastes. Later, the VARTM technique for creating the sample is also briefly discussed. The carpet waste was effectively used to create composite material using the VARTM technology due to its proper infusion and indefectibility. Due to the consistent infusion of resin into the carpet strands, the VARTM process is suited for producing any carpet composite. The carpet waste acts as reinforcement, with the epoxy resin acting as the composite's matrix. The application of the efficient VARTM method proposed a functional composite material development.

References

Alireza, Zareei Seyed, Farshad Ameri, Nasrollah Bahrami, Parham Shoaei, Hamid Reza, Foad Nurian, 2019. "Green High Strength Concrete Containing Recycled Waste Ceramic Aggregates and Waste Carpet Fibers: Mechanical, Durability, and Microstructural Properties." *Journal of Building Engineering* 26: 100914. doi:10.1016/j.jobe.2019.100914.

Alrshoudi, Fahed, Hossein Mohammadhosseini, Mahmood Md Tahir, Rayed Alyousef, Hussam Alghamdi, Yousef R. Alharbi, and Abdulaziz Alsaif. 2020. "Sustainable Use of Waste Polypropylene Fibers and Palm Oil Fuel Ash in the Production of Novel Prepacked Aggregate Fiber-Reinforced Concrete." *Sustainability* 12 (12): 1–14. doi:10.3390/su12124871.

Atakan, Raziye, Serdar Sezer, and Hale Karakas. 2018. "Development of Nonwoven Automotive Carpets Made of Recycled PET Fibers with Improved Abrasion Resistance." *Journal of Industrial Textiles* 49: 835–57. doi:10.1177/1528083718798637.

Awal, A. S. M. Abdul M. Abdul, and Hossein Mohammadhosseini. 2016. "Green Concrete Production Incorporating Waste Carpet Fiber and Palm Oil Fuel Ash." *Journal of Cleaner Production* 137: 157–66. doi:10.1016/j.jclepro.2016.06.162.

Basu, Swapan, and Ajay Kumar Debnath. 2015. "Boiler Control System." In *Power Plant Instrumentation and Control Handbook*, 2nd ed., 585–694. Geneva, Switzerland: Elsevier Ltd. doi:10.1016/b978-0-12-800940-6.00008-3.

Bateman, Stuart A., and Dong Yang Wu. 2001. "Composite Materials Prepared from Waste Textile Fiber." *Journal of Applied Polymer Science* 81 (13): 3178–85. doi:10.1002/app.1770.

Bhatt, Alpa Tapan, Piyush P. Gohil, and Vijaykumar Chaudhary. 2022. "Degassing and Layers Variation Effect on Composite Processing by Vacuum Assisted Resin Transfer Moulding." *Journal of Engineering Research (Kuwait)* 10 (2 B): 184–92. doi:10.36909/jer.9941.

Brebu, Mihai. 2020. "Environmental Degradation of Plastic Composites with Natural Fillers-a Review." *Polymers* 12 (1): 166 (1-22). doi:10.3390/polym12010166.

Das, Sarat. 2009. *"Impact of Nanoclay on Fire Retardancy and Environmental Durability of Post Consumer Carpet Composites."* Bangladesh University of Engineering and Technology Dhaka, Bangladesh. https://shareok.org/bitstream/handle/11244/9917/Das_okstate_0664M_12549.pdf;jsessionid=37515AAE09C95F1E2E9D089E1BE1AE40?sequence=1.

Dayiary, M., S. Shaikhzadeh Najar, and M. Shamsi. 2009. "A New Theoretical Approach to Cut-Pile Carpet Compression Based on Elastic-Stored Bending Energy." *Journal of the Textile Institute* 100 (8): 688–94. doi:10.1080/00405000802170242.

Dubey, Ankit Dhar, Jogendra Kumar, Shivi Kesarwani, and Rajesh Kumar Verma. 2021. "Flexural Properties and Cost Evaluation of Hybrid Polymer Composites Developed from Different Stacking Sequences of Fiber Laminates." *E3S Web of Conferences* 309: 01017. doi:10.1051/e3sconf/202130901017.

Gangineni, Pavan Kumar, Sagar Yandrapu, Sohan Kumar Ghosh, Abhijeet Anand, Rajesh Kumar Prusty, and Bankim Chandra Ray. 2019. "Mechanical Behavior of Graphene Decorated Carbon Fiber Reinforced Polymer Composites: An Assessment of the Influence of Functional Groups." *Composites Part A: Applied Science and Manufacturing* 122: 36–44. doi:10.1016/j.compositesa.2019.04.017.

Gay, Daniel, Suong V. Hoa, and Stephen W. Tsai. 2002. *Composite Materials Design and Applications*. Edited by Daniel Gay, Suong V. Hoa, and Stephen W. Tsai. *Taylor and Francis*. 3rd ed. Boca Raton, Florida: CRC Press. doi:10.1201/9781420031683.

George, Jayamol, M. S. Sreekala, and Sabu Thomas. 2001. "A Review on Interface Modification and Characterization of Natural Fiber Reinforced Plastic Composites." *Polymer Engineering and Science* 41 (9): 1471–85. doi:10.1002/pen.10846.

Gowayed, Yasser, and Mahmoud M. El-halwagi. 1995. "Synthesis of Composite Materials from Waste Fabrics and Plastics." *Journal of Elastomers & Plastics* 27: 79–90. doi:10.1177/009524439502700106.

Haines, Sarah R., Rachel I. Adams, Brandon E. Boor, Thomas A. Bruton, John Downey, Andrea R. Ferro, Elliott Gall, Brett J. Green, Bridget Hegarty, Elliott Horner, David E. Jacobs, Paul Leieux, Powel K. Misztal, Glenn Morrison, Matthew Perzonowski, Tiina Reponen, Rachael E. Rush, Troy Virgo, Celine Alkhayri, Ashleigh Bope, Samuel Cochran, Jennie Cox, Allie Donohue, Andrew A. May, Nicholas Nastasi, Marcia Nishioka, Nicole Renninger, Yilin Tian, Christina Uebel-Niemeier, David Wilkinson, Tianren Wu, Jordan Zambrana, Karen C. Dannemiller. 2020. "Ten Questions Concerning the Implications of Carpet on Indoor Chemistry and Microbiology." *Building and Environment* 170: 106589. doi:10.1016/j.buildenv.2019.106589.

Horoschenkoff, Alexander, and Christian Christner. 2012. "Carbon Fibre Sensor: Theory and Application." In *Composites and Their Applications*, edited by Ning Hu, 3484. Chiba University, Japan: Intech Open. doi:10.5772/50504.

Hosur, Mahesh, Tanjheel Mahdi, and Shaik Jeelani. 2018. "Studies on the Performance of Multi-Phased Carbon/Epoxy Composites with Nanoclay and Multi-Walled Carbon Nanotubes." *Multiscale and Multidisciplinary Modeling, Experiments and Design* 1 (4): 255–68. doi:10.1007/s41939-018-0017-9.

Islam, Shafiqul, and Gajanan Bhat. 2019. "Environmentally-Friendly Thermal and Acoustic Insulation Materials from Recycled Textiles." *Journal of Environmental Management* 251: 109536. doi:10.1016/j.jenvman.2019.109536.

Jain, Abhishek, Gajendra Pandey, Abhishek K. Singh, Vasudevan Rajagopalan, Ranji Vaidyanathan, and Raman P. Singh. 2012. "Fabrication of Structural Composites from Waste Carpet." *Advances in Polymer Technology* 31 (4): 380–89. doi:10.1002/adv.20261.

Jannatyha, Narges, Saeedeh Shojaee-Aliabadi, Maryam Moslehishad, and Ehsan Moradi. 2020. "Comparing Mechanical, Barrier and Antimicrobial Properties of Nanocellulose/CMC and Nanochitosan/CMC Composite Films." *International Journal of Biological Macromolecules* 164: 2323–28. doi:10.1016/j.ijbiomac.2020.07.249.

Kedari, Vishwanath R., Basil I. Farah, and Kuang Ting Hsiao. 2011. "Effects of Vacuum Pressure, Inlet Pressure, and Mold Temperature on the Void Content, Volume Fraction of Polyester/e-Glass Fiber Composites Manufactured with VARTM Process." *Journal of Composite Materials* 45 (26): 2727–42. doi:10.1177/0021998311415442.

Kesarwani, Shivi, Puranjay Pratap, Jogendra Kumar, Rajesh Kumar Verma, and Vijay Kumar Singh. 2021. "An Integrated Approach for Machining Characteristics Optimization of Polymer Nanocomposites." *Materials Today: Proceedings* 44: 2638–44. doi:10.1016/j.matpr.2020.12.672.

Kocak, Dilara, Mehmet Akalin, and Nigar Merdan. 2016. "Investigation of the Mechanical, Thermal, and Morphological Properties of Isotactic Polypropylene/Linear Low-Density Polyethylene/Ethylene Vinyl Acetate Polymer Blends and Their Fibers." *Journal of Industrial Textiles* 45 (5): 879–95. doi:10.1177/1528083714542826.

Koch, Dietmar, Kamen Tushtev, Jürgen Horvath, Ralf Knoche, and Georg Grathwohl. 2006. "Evaluation of Mechanical Properties and Comprehensive Modeling of CMC with Stiff and Weak Matrices." *Advances in Science and Technology* 45: 1435–43. doi:10.4028/www.scientific.net/AST.45.1435.

Kumar, Amit, Kamal Sharma, and Amit Rai Dixit. 2019. "A Review of the Mechanical and Thermal Properties of Graphene and Its Hybrid Polymer Nanocomposites for Structural Applications." *Journal of Materials Science* 54 (8): 5992–6026. doi:10.1007/s10853-018-03244-3.

Kumar, Jogendra, Kumar Abhishek, and Jinyang Xu. 2022. "Experimental Investigation on Machine-Induced Damages during the Milling Test of Graphene/Carbon Incorporated Thermoset Polymer Nanocomposites." *Journal of Composites Science* 6 (77): 1–12. doi:10.3390/jcs6030077.

Kumar, Jogendra, Kaushlendra Kuldeep Kaushlendra Kumar, Balram Jaiswal, Kaushlendra Kuldeep Kaushlendra Kumar, and Rajesh Kumar Verma. 2022. "Investigation on the Physio-Mechanical Properties of Carpet Waste Polymer Composites Incorporated with Multi-Wall Carbon Nanotube (MWCNT)." *Journal of the Textile Institute*, 1–10. doi:10.1080/00405000.2022.2062860.

Kumar, Jogendra, Kuldeep Kaushlendra Kumar, Kuldeep Kaushlendra Kumar, Balram Jaiswal, and Rajesh K. Verma. 2022. "Development of Waste Carpet (Jute) and Multi-Wall Carbon Nanotube Incorporated Epoxy Composites for Lightweight Applications." *Progress in Rubber Plastics and Recycling Technology* 38 (3): 247–63. doi:10.1177/14777606221110252.

Kumar, Jogendra, Rajesh K. Verma, Arpan K. Mondal, and Vijay K. Singh. 2021. "A Hybrid Optimization Technique to Control the Machining Performance of Graphene/Carbon/Polymer (Epoxy) Nanocomposites." *Polymers and Polymer Composites* 29 (9): S1168–80. doi:10.1177/09673911211046789.

Kumar, Jogendra, and Rajesh Kumar Verma. 2021. "Multiple Response Optimization in Machining (Milling) of Graphene Oxide-Doped Epoxy/Cfrp Composite Using CoCoSo-PCA: Anovel Hybridization Approach." *Journal of Advanced Manufacturing Systems* 20 (2): 423–46. doi:10.1142/S0219686721500207.

Kumar, Jogendra, Rajesh Kumar Verma, and Arpan Kumar Mondal. 2020. "Predictive Modeling and Machining Performance Optimization during Drilling of Polymer Nanocomposites Reinforced by Graphene Oxide/Carbon Fiber." *Archive of Mechanical Engineering* 67 (2): 229–58. doi:10.24425/ame.2020.131692.

Kumar, Mohit, J. S. Saini, and H. Bhunia. 2020. "Performance of Mechanical Joints Prepared from Carbon-Fiber-Reinforced Polymer Nanocomposites under Accelerated Environmental Aging." *Journal of Materials Engineering and Performance* 29. Springer US: 7511–7525. doi:10.1007/s11665-020-05216-8.

Kumar, Tejendra, Sanjay Mishra, and Rajesh Kumar Verma. 2019. "Fabrication and Tensile Behavior of Post-Consumer Carpet Waste Structural Composite." *Materials Today: Proceedings* 26: 2216–20. doi:10.1016/j.matpr.2020.02.481.

Kumar Verma, Rajesh, Balram Jaiswal, Rahul Vishwakarma, Kuldeep Kaushlendra Kumar, Kuldeep Kaushlendra Kumar, Rajesh Kumar Verma, Balram Jaiswal, Rahul Vishwakarma, and Kuldeep Kaushlendra Kumar. 2021. "Polymer Composite Developed from Discarded Carpet for Light Weight Structural Applications : Development and Mechanical Analysis." *E3S Web of Conferences* 309: 01154 (1-5). doi:10.1051/e3sconf/202130901154.

Kunrath, Kamila, Eduardo Fischer Kerche, Mirabel Cerqueira Rezende, and Sandro Campos Amico. 2019. "Mechanical, Electrical, and Electromagnetic Properties of Hybrid Graphene/Glass Fiber/Epoxy Composite." *Polymers and Polymer Composites* 27 (5): 1–6. doi:10.1177/0967391119828559.

Liu, De Fu, Yong Jun Tang, and W. L. Cong. 2012. "A Review of Mechanical Drilling for Composite Laminates." *Composite Structures* 94 (4): 1265–79. doi:10.1016/j.compstruct.2011.11.024.

Miraftab, M., and M. Mirzababaei. 2009. "Carpet Waste Utilisation, an Awakening Realisation: A Review." In *Second International Symposium on Fibre Recycling*, 1–9. Georgia Institute of Technology, Atlanta, Georgia, USA. doi:10.13140/RG.2.1.3880.3366.

Miraftab, Mohsen, Richard Horrocks, and Colin Woods. 1999. "Carpet Waste, an Expensive Luxury We Must Do Without." *Autex Research Journal* 1 (1): 1–7. doi:10.1533/9780857092991.3.173.

Mishra, Kunal, Sarat Das, Ranji Vaidyanathan, and Tooling Materials. 2019. "The Use of Recycled Carpet in Low-Cost Composite Tooling Materials." *Recycling* 4: 12 (1-8). doi:10.3390/recycling4010012.

Mohammed, Layth, M. N.M. Ansari, Grace Pua, Mohammad Jawaid, and M. Saiful Islam. 2015. "A Review on Natural Fiber Reinforced Polymer Composite and Its Applications." *International Journal of Polymer Science* 2015: 243947 (1-15). doi:10.1155/2015/243947.

Myagkov, L. L., K. Mahkamov, N. D. Chainov, and I. Makhkamova. 2014. "Advanced and Conventional Internal Combustion Engine Materials." In *Alternative Fuels and Advanced Vehicle Technologies for Improved Environmental Performance: Towards Zero Carbon Transportation*, edited by Richard Folkson, 370–92. Cambridge, UK: Woodhead Publishing Limited. doi:10.1533/9780857097422.2.370 Abstract:

Naebe, Minoo, Jing Wang, Abbas Amini, Hamid Khayyam, Nishar Hameed, Lu Hua Li, Ying Chen, and Bronwyn Fox. 2014. "Mechanical Property and Structure of Covalent Functionalised Graphene/Epoxy Nanocomposites." *Scientific Reports* 4: 4375 (1-7). doi:10.1038/srep04375.

Nguyen, Tuan Anh, Quang Tung Nguyen, and Trong Phuc Bach. 2019. "Mechanical Properties and Flame Retardancy of Epoxy Resin/Nanoclay/Multiwalled Carbon Nanotube Nanocomposites." *Journal of Chemistry* 2019: 3105205 (1-9). doi:10.1155/2019/3105205.

Nijssen, R.P.L., and R.P.L.Nijssen. 2015. *Composite Materials an Introduction*. Edited by R. P. L. Nijssen. *A VKCN Publication*. 2015th ed. Vol. 1. Inholland University of Applied Sciences. www.inholland.nl/lectoraatgrootcomposiet.

Nonobe, Yasuhiro. 2017. "Development of the Fuel Cell Vehicle Mirai." *IEEJ Transactions on Electrical and Electronic Engineering* 12 (1): 5–9. doi:10.1002/tee.22328.

Onal, L., and Y. Karaduman. 2009. "Mechanical Characterization of Carpet Waste Natural Fiber-Reinforced Polymer Composites." *Journal of Composite Materials* 43 (16): 1751–68. doi:10.1177/0021998309339635.

Papageorgiou, Dimitrios G., Mufeng Liu, Zheling Li, Cristina Vallés, Robert J. Young, and Ian A. Kinloch. 2019. "Hybrid Poly(Ether Ether Ketone) Composites Reinforced with a Combination of Carbon Fibres and Graphene Nanoplatelets." *Composites Science and Technology* 175: 60–68. https://doi.org/10.1016/j.compscitech.2019.03.006.

Pathak, Abhishek K., Hema Garg, Mandeep Singh, T. Yokozeki, and Sanjay R. Dhakate. 2019. "Enhanced Interfacial Properties of Graphene Oxide Incorporated Carbon Fiber Reinforced Epoxy Nanocomposite: A Systematic Thermal Properties Investigation." *Journal of Polymer Research* 26 (2): 23 (1-23). doi:10.1007/s10965-018-1668-2.

Qin, Q. H. 2015. *Introduction to the Composite and Its Toughening Mechanisms*. Edited by Qinghua Qin and Jianqiao Ye. *Toughening Mechanisms in Composite Materials*. Cambridge, UK: Woodhead Publishing. doi:10.1016/B978-1-78242-279-2.00001-9.

Rubino, Felice, Antonio Nisticò, Fausto Tucci, and Pierpaolo Carlone. 2020. "Marine Application of Fiber Reinforced Composites: A Review." *Journal of Marine Science and Engineering* 8 (1): 26 (1-28). doi:10.3390/jmse8010026.

Rushforth, I. M., K. V. Horoshenkov, M. Miraftab, and M. J. Swift. 2005. "Impact Sound Insulation and Viscoelastic Properties of Underlay Manufactured from Recycled Carpet Waste." *Applied Acoustics* 66: 731–49. doi:10.1016/j.apacoust.2004.10.005.

Shagor, R. M.R., F. Abedin, and R. Asmatulu. 2021. "Mechanical and Thermal Properties of Carbon Fiber Reinforced Composite with Silanized Graphene as Nano-Inclusions." *Journal of Composite Materials* 55 (5): 597–608. doi:10.1177/0021998320953183.

Shin, Pyeong Su, Dong Jun Kwon, Jong Hyun Kim, Sang Il Lee, K. Lawrence DeVries, and Joung Man Park. 2017. "Interfacial Properties and Water Resistance of Epoxy and CNT-Epoxy Adhesives on GFRP Composites." *Composites Science and Technology* 142: 98–106. doi:10.1016/j.compscitech.2017.01.026.

Singh, Kulvir. 2014. "Advanced Materials for Land Based Gas Turbines." *Transactions of the Indian Institute of Metals* 67 (5): 601–15. doi:10.1007/s12666-014-0398-3.

Siwal, Samarjeet Singh, Qibo Zhang, Nishu Devi, and Vijay Kumar Thakur. 2020. "Carbon-Based Polymer Nanocomposite for High-Performance Energy Storage Applications." *Polymers* 12 (3): 505 (1-31). doi:10.3390/polym12030505.

Sohn, Y. K. 1997. "On Traction-Carpet Sedimentation." *Journal of Sedimentary Research* 67 (3): 502–9. doi:10.1306/D42685AE-2B26-11D7-8648000102C1865D.

Sotayo, Adeayo, Sarah Green, and Geoffrey Turvey. 2018. "Development, Characterisation and Finite Element Modelling of Novel Waste Carpet Composites for Structural

Applications." *Journal of Cleaner Production* 183: 686–97. doi:10.1016/j.jclepro.2018.02.095.
Stewart, Richard. 2009. "Lightweighting the Automotive Market." *Reinforced Plastics* 53 (3): 14–21. doi:10.1016/S0034-3617(09)70078-5.
Street, Carrie, Justin Woody, Jaime Ardila, and Miguel Bagajewicz. 2008. "Product Design: A Case Study of Slow-Release Carpet Deodorizers/Disinfectants." *Industrial & Engineering Chemistry Research* 47 (4). American Chemical Society: 1192–1200. doi:10.1021/ie0710622.
Thakur, Vijay Kumar, and Manju Kumari Thakur. 2014. "Processing and Characterization of Natural Cellulose Fibers/Thermoset Polymer Composites." *Carbohydrate Polymers* 109: 102–17. doi:10.1016/j.carbpol.2014.03.039.
Uygunoglu, Tayfun, Ibrahim Gunes, and Witold Brostow. 2015. "Physical and Mechanical Properties of Polymer Composites with High Content of Wastes Including Boron." *Materials Research* 18 (6): 1188–96. doi:10.1590/1516-1439.009815.
Vijay, N., V. Rajkumara, and P. Bhattacharjee. 2016. "Assessment of Composite Waste Disposal in Aerospace Industries." *Procedia Environmental Sciences* 35: 563–70. doi:10.1016/j.proenv.2016.07.041.
Wang, Ru-Min, Shui-Rong Zheng, and Ya-Ping Zheng. 2011. "Introduction to Polymer Matrix Composites." In *Polymer Matrix Composites and Technology*, edited by Ru-Min Wang, Shui-Rong Zheng, and Ya-Ping Zheng, 1–25. Cambridge, UK: Woodhead Publishing Limited. doi:10.1533/9780857092229.1.
Wang, Youjiang. 2006. *Recycling in Textiles*. Edited by Youjiang Wang. *Woodhead Publishing*. Cambridge England: Woodhead publishing limited. doi:10.1533/9781845691424.1.
Zhao, Yun-Hong Hong, Ya-Fei Fei Zhang, Shu-Lin Lin Bai, and Xiao-Wen Wen Yuan. 2016. "Carbon Fibre/Graphene Foam/Polymer Composites with Enhanced Mechanical and Thermal Properties." *Composites Part B: Engineering* 94: 102–8. doi:10.1016/j.compositesb.2016.03.056.

Chapter 3

Mechanical Properties of Composites Developed from Carpet Waste

This chapter intends to present the basic idea of mechanical characterization and performance assessment of composites created from carpet waste. The proposed material would emphasize the significance of the ongoing research efforts and their intended uses. It offers the necessary understanding of the properties of composites made from recycled waste carpets and their scientific application. The analysis setup for a carpet waste/epoxy composite has been covered in this chapter. Based on the previous research investigation, it is largely focusing on the mechanical properties (tensile, flexural, and impact strength) of waste carpet composites.

Mechanical Performance Investigation and Characterization

A variety of polymer groups are combined to achieve a wide range of functions. This is expected to lead to the development of new components with improved qualities. However, alterations in physical attributes (ILSS, IFSS, and tensile strength) have proven suitable for industrial applications (A. Kumar, Sharma, and Dixit 2019; Adak et al. 2018). Because of this, several forms of reinforcement were utilized as an addition to the composites. These reinforcements possessed higher mechanical strength and stiffness levels than the polymer materials (Fan et al. 2010). Compared to metals and their alloys, the strength-to-weight ratio of reinforced polymer composites may be viewed and measured more reliably. It is common knowledge that the introduction of fibers may significantly boost a material's mechanical qualities; nevertheless, the production cost and complexity of the product will rise as a result of this type of processing. The use of fiber as an additive makes it possible to make advancements without significantly increasing the weight of the composite is an essential facet of this strategy (Burmistrov et al. 2016; Coleman et al. 2006; Choudhary and Gupta 2001). Researchers have developed a number of strategies to expand the mechanical properties of fabrics, including fiber seaming and 3D fabric design, as well as surface fiber and matrix changes. Polymers are easier to work with and more cost-effective than metals when modifying mechanical characteristics in the matrix phase (Pathak et al. 2016). Fiber orientation, fiber layer, fiber type, and fiber volume % all have a role in composite mechanical performance. Therefore, adding fibers greatly influences polymer composites' mechanical and other properties. Several interface theories, such as chemical bonding mechanism, wetting mode, mechanical interlocking, and polymer matrix local stiffness, can explain the combined impact of the alteration. It is targeted mainly to improve the polymer's mechanical and physical properties with the addition of reinforcement such as carbon fiber or Kevlar fiber, etc. (Domun et al. 2015; Pal, Sharma, and Sharma 2014). In today's composite production, fiber may readily substitute other common and nonconventional additives to achieve excellent mechanical properties (Ghorai 2013).

Furthermore, managing and treating waste carpet fiber in the composite matrix is challenging from an engineering standpoint. As a result, obtaining homogeneous characteristics across the composite matrix requires stronger interphase interaction between waste carpet and matrix (Phiri, Gane, and Maloney 2017; Naebe et al. 2014). This type of complex issue can be solved with the use of a polymer matrix. Using different fillers and reinforcement materials, a polymer composite could continually enhance matrix materials' qualities. The contact area between the reinforcement and the polymer matrix is large per volume because of the connectivity. As a result of these synergistic effects (self-healing, ternary system, morphological modification, and carrier mobility) in the matrix, aspect ratios are improved. Mechanically strong materials are ideal for multifunctional items because of their durability and stiffness (Zhang et al. 2015).

In the literature work of eminent scholars, it has been reported to intend the discarded fiber materials (M. S. S. Kumar et al. 2014; Chaudhery and Hussain 2019). Various extensive surveys were proposed to evaluate fiber feasibility and its role in manufacturing superior polymer composites. The high specific modulus and strength of CFRP and GFRP (CF/GF) have attracted a lot of interest in research and innovation. It is extensively employed in a wide range of high-performance, lightweight structural components (Stewart 2009). However, CFRP composite laminates are extremely sensitive to the beginning transmission of cracks and failures and fiber/matrix interfaces because they are constructed in the form of a laminated structure. As a result, the research and development of innovative strategies to improve the CF/GF reinforced polymer interlaminar strength are necessary.

Several approaches for improving interlaminar characteristics and damage control in CF/GF laminates have been presented in the literature. The matrix phase for fiber-reinforced composite materials was made with epoxy resin. It has been widely utilized in laminate adhesives and coatings due to its excellent mechanical, electrical, abrasion, and chemical reaction resistance qualities. Fiber-reinforced polymer composites have extraordinary properties such as enhanced strength, durability, reduced weight, thermal stability, corrosion resistance, stiffness, and so on. The manufacturing industry sectors widely recognize these essential features in a variety of applications, including turbine blades, sporting goods, aviation, marine components, and automobiles (Gang 2018; Jiménez-Pérez, Kharissova, and Flores 2017; Sukanchan Palit and Mustansar Hussain 2018). Other advantages over other raw materials include its inexpensive cost of production as well as its custom size, surface polish, and ease of handling.

In the case of plastics manufacture, the fibers serve a significant role in capacitating the load-bearing function and transmitting this load to the matrix phase. The fiber-reinforced content primarily determines the mechanical characteristics of the composite. Flexural and laminar shear strength qualities are controlled by fiber meshing (Chen et al. 2019). The fiber and matrix bonding in multilayer carbon/polymer composites is more prone to breaking or spreading through multiple failure mechanisms. Another major concern connected to breaking mechanisms is degradation, resulting in lower tensile strength and stiffness and a shorter product life cycle. Moreover, fiber surface inertness is an essential factor that might affect polymer matrix bonding. The inappropriate mixing at the interphase of two distinct matrices results in substantial damage and failure of the whole composite assembly (Yuan et al. 2019).

Carbon-fiber reinforced composites (CFRCs) have replaced their metal counterparts with their preferred strength-to-weight and rigidity-to-weight ratios for a wide range of high-performance structural purposes. In order to provide an effective transfer of load from the matrix to the fillers, the interface parameters of the fiber matrix determine the strength of fiber/polymers. To minimize stress and improve mechanical performance, interfacial properties are essential. Innovations in carbon fiber/epoxy matrix strength are becoming an extensive research area for trade and society (Gang 2018; Ning et al. 2015).

Waste re-utilization provides a feasible way to develop cost-effective products and reduce environmental hazards. Very limited studies exist on reusing carpets for composite development and waste management. Nowadays, numerous categories of carpet waste exist in garbage piles that are problematic to recycle. Several pioneering experiments have been conducted to utilize such waste appropriately. An increase in the proper utilization of this waste can eliminate its concerns. Continuous research has been undertaken to manage such wastes and provide an alternate solution.

For sustainable products, materials recovery and recycling must be based on cost-effective technology and beneficial environmental impact throughout their life cycle. The carpet is made of petrochemical-based materials; recycling carpet may save around 700,000 barrels of oil each year, resulting in total energy conservation of 4.4 trillion British thermal units (BTUs) (Mishra and Vaidyanathan 2019b; Mihut et al. 2001). Discarded carpet becomes a massive waste and is harmful to the environment and the economy when the users throw it out in open areas as it involves cost issues during decomposition and landfilling after its application. It requires the cost of dumping valuable raw materials like nylon 6,6, polyester, polyvinyl chloride, polypropylene, and olefins. The landfilling of post-consumer carpet fibers having non-biodegradable nature is currently producing several environmental problems (Jain et al. 2012; Mihut et al. 2001). One sector with large waste is the textile sector, though these wastes contain valuable fiber products. These wastes can be recycled to make various products for several applications, including thermal and acoustic applications (Alhuthali, Low, and Dong 2012). This recycling process is beneficial for reducing waste and fighting against severe environmental issues. Many researchers investigate the method to recycle the waste generated from the textile and carpet industries into a useful form. Wang et al. (Wang, Ucar, and Wang 2014) developed the lightweight cemented board by utilizing recycled post-consumer carpet waste fiber. The findings revealed that this approach could reduce the landfilling of the waste carpet and lead to a pollution-free environment. Mohammadhosseini et al. (Mohammadhosseini, Awal, and Mohd 2017) studied the potential use of waste polypropylene fiber and exhibited concrete's impact resistance and mechanical characteristics with fiber. They observed that the addition of waste carpet enhanced fiber-reinforced concrete's resistance and energy absorption capacity. The authors concluded that the developed composite materials could be recommended for building construction applications. Islam et al. (Islam and Bhat 2019) fabricated an environment-friendly structural composite from discarded waste textile materials for acoustic and thermal applications. The work concluded that these developed composite materials could play a vital role in energy savings and reducing the environmental pollution. Alrshoudi et al. (Alrshoudi et al. 2020) experimented with ways to use waste in buildings. According to the researchers, there has been a positive correlation between polypropene carpets and palm oil fuel ash. In order to improve flexural and tensile strength, elasticity, energy absorption, and impact resistance of the matrix, the fibers were included in the design. Natural fiber composites have gained popularity because of their low density, affordability, and certain suitable qualities, among other things. Due to growing environmental awareness, hemp, flax, and jute fibers have attracted a lot of attention as composite reinforcing materials recently. These are great options because they are strong, affordable, renewable, and lightweight. Due to its low cost and mass production, jute holds a distinct position in the market. Jute fiber has exceptional properties, including high specific strength and modulus. Jute yarns and textiles have recently been widely used in the textile sector as packaging and carpet backing. Jute fibers are often used as weft threads in carpets, where they are produced in significant amounts. If the weft thread used in the carpet-making process is longer than is necessary, it has been found that this will happen. Before placing the carpet on the floor, it is necessary to trim the excess uneven weft yarn ends in order to get smooth and level carpet sides 5-7% of jute weft threads are thought to be cut and eliminated as trash during the carpet-making process (Onal and Karaduman 2009). As a result, the issue of waste carpet recycling is inspired by carpet recycling in structural composite products. Awal et al. (Awal and Mohammadhosseini 2016) developed a unique carpet fiber reinforced concrete composite out of polypropylene waste carpet fiber and ordinary concrete. It was discovered that putting carpet fiber into concrete reduced the workability and density substantially. It increased flexural and tensile strength. A relationship between decreasing modulus of elasticity and fiber quantity was discovered. According to the author, the shrinking strain of manufactured composite was also larger than that of regular plain concrete, which improved soil cohesion and reinforcement. Jain et al. (Jain et al. 2012) utilized the vacuum-assisted resin transfer molding (VARTM) technique to develop the engineering composite from post-consumer carpet waste. Two types of carpets, such as nylon and olefin-based, were used for fabrication. They concluded that composites could make various feasible applications, including low-cost containers, civil infrastructure components, materials for impact protection, etc.

For packaging and protective applications, reducing fire and water uptake is essential. Miraftab et al. (Miraftab, Horrocks, and Woods 1999) experimented with nylon carpet fiber in clay to study green and sustainable development. This combination showed superior load-bearing, internal cohesion, shear, and comprehensive strength compared to conventional clay or subpar soils. A high soil moisture content greatly aids the qualities as mentioned above. Fiber type, fiber length, soil nature, and environmental conditions can also affect the results of such prepared composite to augment the mechanical performance of glass fiber-reinforced polymer (GFRP) properties. Mishra et al. (Mishra and Vaidyanathan 2019a) studied the mechanical properties and fire retardant of developed composite by using recycled waste carpet. For the development of nanocomposite, the VARTM technique was used. The flame retardancy results revealed that the presence of the clay in the developed sample decreases the ignition time. Choobbasti et al. (Choobbasti, Samakoosh, and Kutanaei 2019) improved the soil's mechanical strength by reinforcing carpet waste fiber and nano calcium carbonate materials. The researcher revealed that the combined nanoparticle and carpet waste fiber significantly improve soil properties.

Materials recovery and recycling can provide viable goods with the requisite cost-competitiveness and environmental aspects for long-term industry growth. Ecological and economic concerns are exacerbated by the amount of carpet and textile industry trash created. This is mainly owing to the expensive cost of carpet removal and pollution caused by burning, landfilling, and simply leaving them out in the open (Teli 2018; Mohammadhosseini and Yatim 2017; dos Reis 2009). The cost includes discarding critical raw materials such as nylon 6,6, polyester, polyvinyl chloride (PVC), polypropylene (PP), and other olefins, in addition to the cost of transporting rubbish to landfills. Because carpet fibers are non-biodegradable, using them as a landfill material causes various environmental issues, including soil pollution. Carpet trash recycling can help with both solid waste management and environmental concerns (Yalcin-enis, Kucukali-ozturk, and Sezgin 2019; Islam and Bhat 2019). The most common fibers in today's mass-produced carpets are olefin, polyester, cotton, and wool. The carpet's face fiber, backing, binder, and filler are all made up of these fibers. Natural fiber composites are becoming increasingly popular because of their low density, low cost, and high quality. Lignocellulosic fibers as composite reinforcement have lately gained great interest due to increased environmental concerns (Onal and Karaduman 2009). It is long-lasting, low-cost, renewable, thin, and gentle on the environment. Jute sets itself apart from other commodities as a low-cost and abundant product. In terms of specific strength and modulus, jute fiber is a better choice.

In recent years, jute yarns and other jute-related goods have become increasingly popular as textile backing materials (Jain et al. 2012; Onal and Karaduman 2009). Weft yarns made from jute fibers are commonly employed in carpets, and the length of these weft threads is generally longer than the stipulated size. Before the carpet is pulled up, the excess uneven weft yarn ends must be clipped to have smooth and consistent carpet sides. According to research, jute weft threads are wasted by 5–7% (Onal and Karaduman 2009). These wastes are recycled to make exceptionally thick yarns or scrap filler for nonwovens. Waste materials have been recycled into valuable items that have desirable characteristics.

Mechanical characterization mainly investigates the material's properties under static and dynamic force conditions. Their testing results ensure that material capacity and suitability for feasible mechanical applications. The outcomes could affect the mechanical strength and overall efficiency of the proposed material to be molded in an appropriate shape and size. It also includes the physical property evaluation of functional characteristics of structural composite material. Herein the focus is made to highlight the mechanical performances of composite samples developed from the discarded carpet under testing modes of the tensile, flexural, impact, compressive, and dynamic mechanical analysis. The findings were explored to understand composites' nature and their end uses.

Tensile Test

The findings of the composites developed from the waste carpet underwent tensile tests such as strength, modulus, and elongation at break. The load cell detects the specimen and resists elongation during the test. The sample preparation and testing conditions significantly impact the tensile strength value. The tensile properties of all materials are determined from the uniaxial tension and specimen prepared according to the ASTM standards. Figure 3.1 (a) shows the dimensions specified per ASTM D3039 standard for carpet/epoxy composite tensile specimens. The setup of a Universal Testing Machine (UTM-WDW-5) was used to examine the prepared samples.

The tensile specimen is placed between the grips of a UTM, as revealed in Figure 3.1 (b), and the load is given gradually until the specimen fails. The average test speed for testing standard test specimens is 2 mm/min. The universal testing equipment is connected to a computer, and experimental data are acquired using software that calculates deformation and tensile strength. Testing in multiple orientations is evaluated depending on the reinforcement and type of composites. This test was repeated for all the specimens chosen for the test. Figure 3.2 (a-d) displays the tensile test samples before testing, and Figure 3.3 (a-d) shows the tensile test samples images after testing.

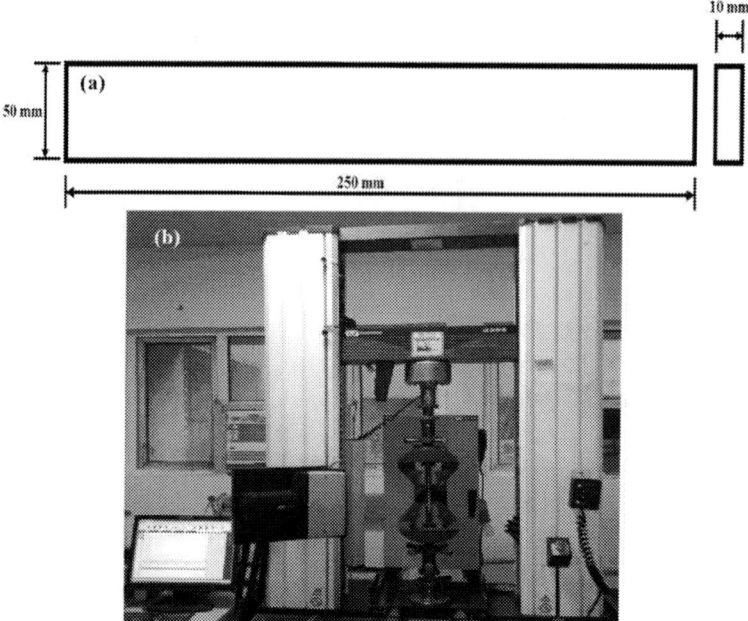

Figure 3.1. (a) Specimen of the tensile test (ASTM D3039), (b) Tensile testing on UTM (Kumar Verma et al. 2021).

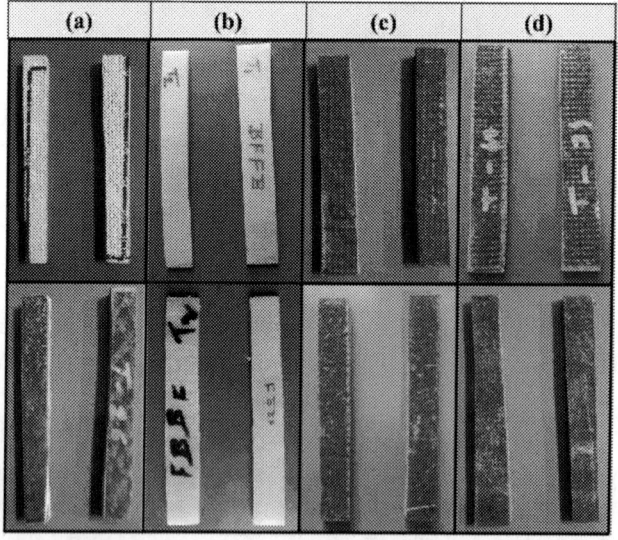

Figure 3.2. Samples of tensile test (a) Wool (b) Polypropylene (c) Nylon (d) Polyester.

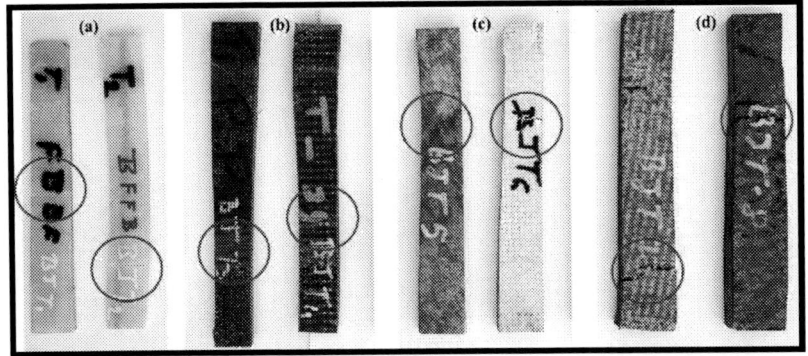

Figure 3.3. Tested samples (tensile) (a) wool (b) polypropylene (c) nylon (d) polyester.

Tensile tests were performed on carpet waste composite materials of wool, polypropylene, nylon, and polyester. All the composite materials were laminated using the two ways back front-front back (BFFB) and front back-back front (FBBF) techniques. The tensile strength of the specimens is revealed in Table 3.1 and plotted stress-strain curves are shown in Figure 3.4 (a-d).

The stress-strain plots show that for wool (FBBF), composite stress increases almost linearly with the strain before the fracture of the specimen. It shows the brittle nature of the wool FBBF composite. In the same way, the wool (BFFB) composite shows a linear rise in the curve but with small yielding before the fracture point. The tensile strength of the wool composite was found 18.20 MPa in FBBF laminate and 16.40 MPa in BFFB laminate. It can be observed that the wool composite strength of FBBF laminate is greater than BFFB laminate.

The tensile strength of the polypropylene composite was determined to be 8 MPa for the FBBF laminate and 7.7 MPa for the BFFB laminate. The stress-strain curve of polypropylene composite shows several peaks without fracture for both laminate FBBF and BFFB, which present the toughness of the composite. It was observed for the polypropylene composite, the strength of FBBF laminate is almost equal to BFFB laminate.

The tensile strength of the nylon composite was found at 7.30 MPa in FBBF laminate and 11.50 MPa in BFFB laminate. The stress-strain curve of the nylon composite rises linearly up to the elastic limit. After that, it shows a yield point and then rises linearly with strain. It presents the ductile nature of nylon composite. The strength of the FBBF laminate for the nylon composite was observed to be smaller than the BFFB laminate.

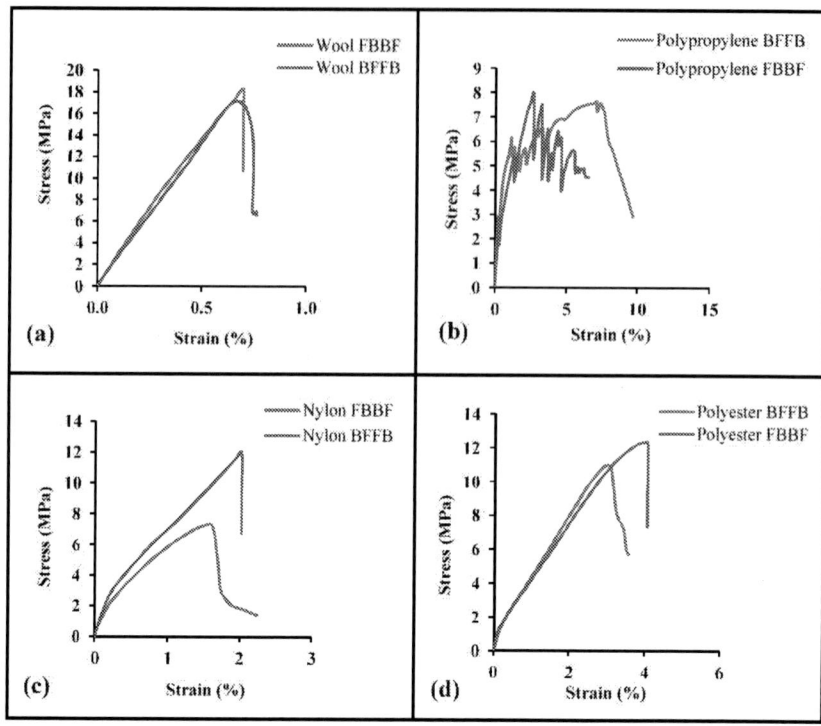

Figure 3.4. Stress-strain (tensile) curves of epoxy-based carpet waste composite (a) Wool (b) Polypropylene (c) Nylon (d) Polyester.

For the polyester composite, both curves of BFFB and FBBF show very low elastic limits. After that, stress rises linearly with strain. The curves present the ductile nature of polyester composite material. Tensile strength was found at 11 MPa in FBBF laminate and 12 MPa in BFFB laminate. The polyester composite's strength of FBBF laminate was observed to be smaller than the BFFB laminate.

The incorporation of waste carpet into the epoxy matrix led to considerable enhancements in the mechanical and physical characteristics of the polymer matrix. The same trend was remarked in the investigation of prior work, which supports the findings of the current work (Rafiee et al. 2019; Pathak et al. 2016). The processing methods, type of carpet materials, and fiber configuration are specific factors affecting properties. As shown from current studies, finding a processing method that is effective in one area but ineffective in another is complex. Performance improvement is a well-known illustration of this, as it generally boosts mechanical properties. As a result, optimizing the various circumstances is essential if the intended composite characteristics are to be achieved. Some studies were carried out to examine the mechanical characteristics of the discarded/polymer composite materials. This shows the feasibility of carpet waste-based composites for mechanical application. It could be an effective solution for solid waste management in the carpet and textile sector using waste material for wealth generation.

Flexural Test

Flexural testing determines a sample's flexural stiffness and bending strength by testing its ability to withstand 3-point and 4-point bending conditions on a UTM machine setup. Flexural strength is defined as the material's ability to resist the bending action against an applied force vertical to the longitudinal axis. The generated stress is a combination of tensile and compressive stress. Flexural strength is also known as rupture modulus, bending, or fracture strength to resist deformation when the sample is being loaded. The bending test in transverse is most frequently used where a specimen with either a rectangular or circular cross-section is bent until fracture. The ASTM D790 standard is commonly used in the polymer and composite sectors and is often quoted for bending strength measurement. The obtained flexural strength of the material allows users to select materials that do not fail when supporting the loads required for the application.

A three-point bend fixture conducted flexural tests on the universal testing machine, as shown in Figure 3.5 (a). The UTM was connected to a computer to display the experimental values on the screen. The flexural strength and the flexural modulus were obtained directly from the computer.

The dimension of the specimens was considered according to ASTM D790 standards, as shown in Figure 3.5 (b). Adjust the support span length with the standard determined length. It places the test specimen on the three-point bend fixture and begins the test on the universal testing machine by applying the load gradually. The bending load was applied to the specimen until it delaminated and broke. This delamination indicates the failure of the sample. Figure 3.6 (a-d) shows the flexural test samples before testing, and Figure 3.7 (a-d) displays the flexural test samples after testing.

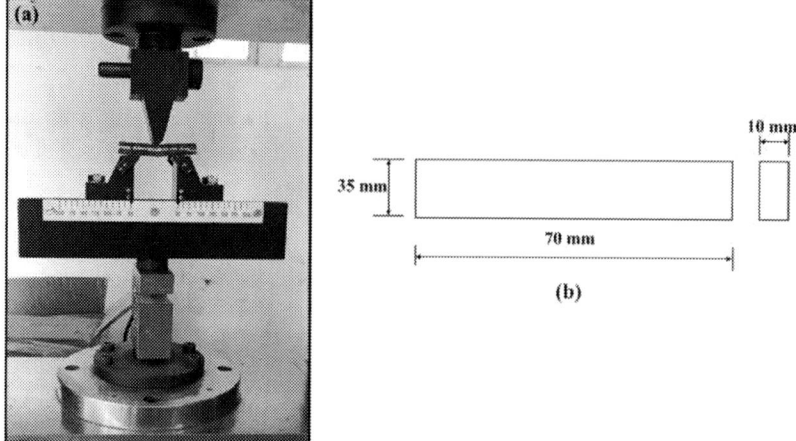

Figure 3.5. (a) Flexural test specimen on UTM, (b) Specimen of the flexural test (ASTM D790).

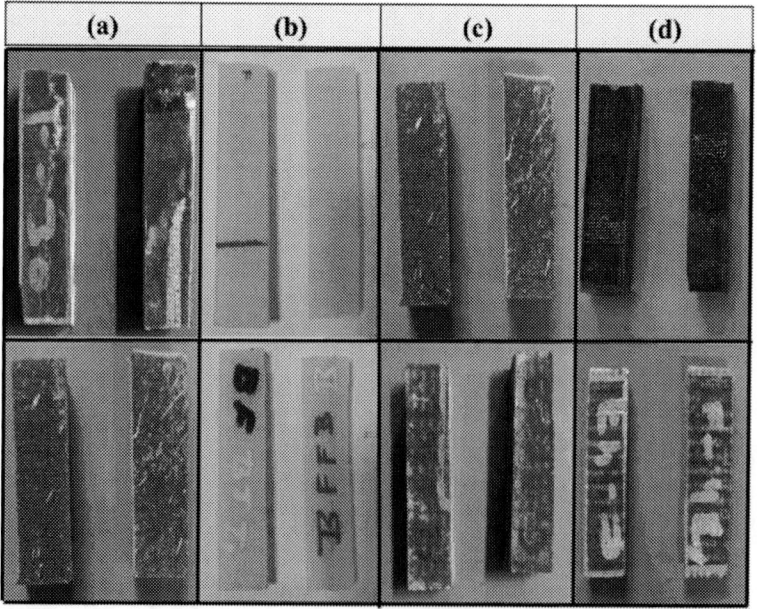

Figure 3.6. Samples of flexural test (a) Wool (b) Polypropylene (c) Nylon (d) Polyester.

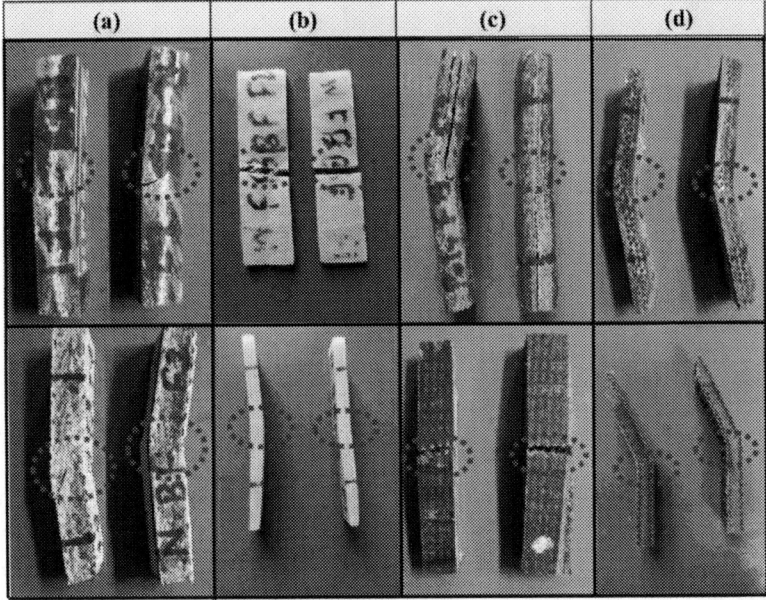

Figure 3.7. Tested samples of flexural test (a) Wool (b) Polypropylene (c) Nylon (d) Polyester.

Flexural strength is defined as the material's ability to resist bending action against an applied force vertically to the longitudinal axis. Flexural tests were done on composites made from various waste carpet materials such as nylon, polypropylene, polyester, and wool. The flexural test is performed to evaluate the specimen's bending strength and flexural stiffness by a 3-point bending test. The stress mode generated during the flexural test comprises tensile and compressive stress. The ASTM D790 standard was followed in preparing the sample. All composite materials were laminated using the two-way back front-front back (BFFB) and front back-back front (FBBF) techniques (FBBF). The flexural strength of discarded carpet/epoxy composite material is shown in Table 3.1. According to the table, the flexural properties of wool/epoxy composite BFFB laminates have the highest flexural strength. Polypropylene (BFFB) and nylon (FBBF) composites have the best flexural strength after wool (BFFB).

In comparison, wool (FBBF) has the lowest flexural strength of the developed structural composite material. The enhancement in flexural properties is clearly due to the recycled waste carpet's ability to withstand the composites' bending force. The polyester/epoxy composite materials BFFB and FBBF, both laminate, were subjected to resistance with the highest load.

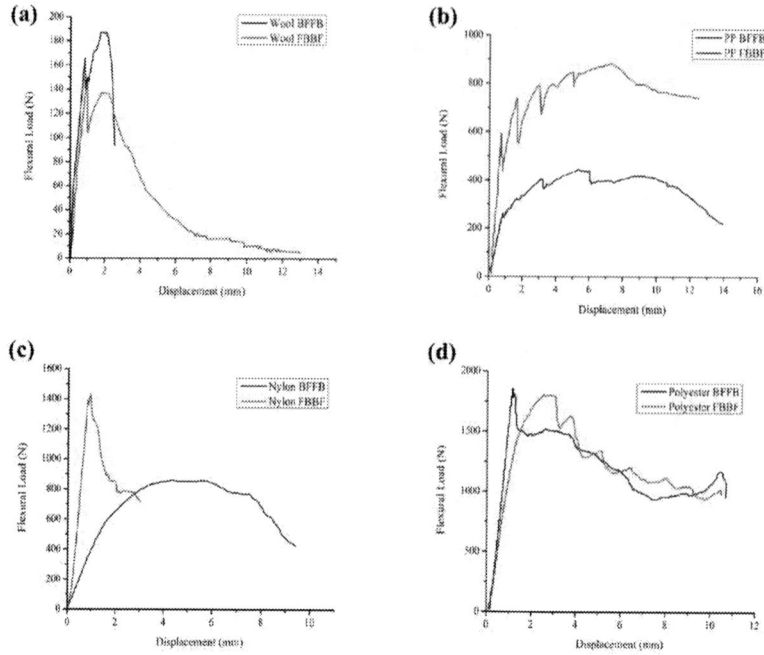

Figure 3.8. Load-displacement (flexural) curves of epoxy-based carpet waste composite (a) Wool (b) Polypropylene (c) Nylon (d) Polyester.

Figure 3.8 demonstrates the typical load-displacement curves attained after performing the flexural test on the four samples. Each graph contains two curves for the same material but different (BFFB & FBBF) configurations. In Figure 3.8 (a), the wool/epoxy BFFB composite exhibits a quick and sharp load rise, indicating a higher load (strength) than the wool/epoxy FBBF sample, but it also has low displacement suggesting that it is brittle. On the other hand, the maximum load of the wool/epoxy FBBF sample is less than the BFFB sample, but the yield displacement is high, showing this composite's ductile nature. Figure 3.8 (b) shows the polypropylene carpet/epoxy composite load-displacement curves. In this graph polypropylene/epoxy FBBF sample has a higher flexural load and maximum yield displacement. This shows the toughness of the material. However, the polypropylene carpet/epoxy BFFB sample has a small rise in load with higher displacement and exhibits the material's ductile property. Figure 3.8 (c) shows the nylon/epoxy composite load-displacement curves. The nylon FBBF sample has a steep load rise with small displacement showing the composite's brittle nature. The nylon BFFB sample, on the other hand, exhibits a moderate rise in load with considerable displacement, indicating the material's ductile nature. Figure 3.8 (d) describes the polyester/epoxy composite load-displacement curves. Both the BFFB and FBBF curves exhibit a sharp rise in load until they reach the elastic limit, after which they display displacement with distinct peaks that demonstrate the materials' toughness. The strength of the composite is determined by the waste that is eliminated during the matrix phase. In terms of load-carrying capability, polyester fiber composite was superior to carbon composite.

Because of the superior mechanical characteristics of the fiber, incorporating one into an epoxy phase matrix should result in structural materials with a much higher modulus and strength. Polymer matrix mechanical characteristics were improved by incorporating rejected fibers. With regard to its properties, a composite material's fiber arrangement has a significant impact. When it comes to polymers, fiber has improved mechanical characteristics, but further investigation of polymer arrangement is still in its infancy. In order to make full use of its unique features in polymer composites, the industrial industry, academia, and research institutions all need to step up their efforts. Many research studies have shown that the mechanical efficiency of composites is due to the poor adhesion between the matrix and waste fiber. The interfacial connection is critical for the composite to have dynamic mechanical properties since load transfers from a matrix to reinforcement fiber are required. The interface shear stress determines how much load may be transferred from the matrix to the reinforcement. To achieve low interfacial shear stress, reinforcement with high interfacial shear stress can transfer the functional load over a long distance.

Table 3.1. Tensile strength of composite material and Flexural strength of composite material

Material	Tensile strength (MPa)		Flexural strength (MPa)	
	FBBF	BFFB	FBBF	BFFB
Wool	18.2	16.4	23.00	34.40
Polypropylene	8	7.7	30.64	34.00
Nylon	7.3	11.5	33.19	25.47
Polyester	11	12	28.23	26.61

Impact Test

The impact test examines the amount of energy absorbed by the composite specimens during fracture mode. This test reveals the material's characteristics as well as the composite's ductility or brittleness performance. The Charpy V-test was conducted in accordance with ASTM D256 standards using a pendulum impact tester V machine and a sample size of 64 mm × 13 mm × 10 mm. The impact testing equipment and breakage sample are shown in Figure 3.9. The average results for each test were computed for two components.

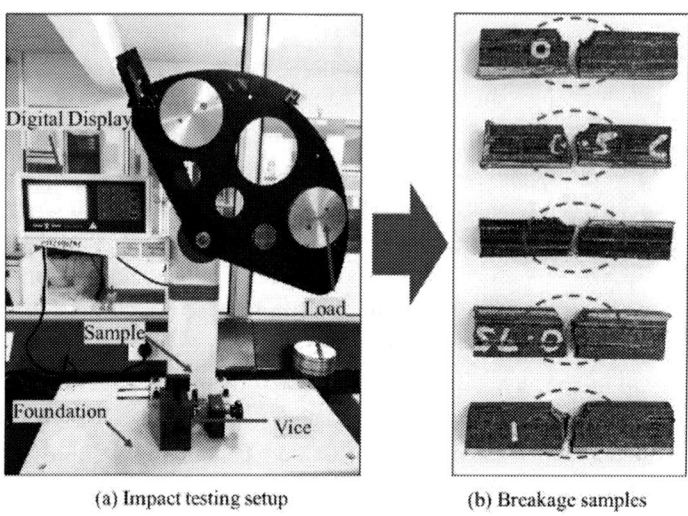

(a) Impact testing setup   (b) Breakage samples

Figure 3.9. Impact test setup and breakage sample (J. Kumar, Kumar, Kumar, et al. 2022).

In order to better understand how a material responds to abrupt forces, the impact test is utilized. This test was carried out in accordance to ASTM D256. The energy and impact strength were measured using Eq. 3.1.

$$I_s = \frac{w \times 10^3}{t \times b_n} \qquad (3.1)$$

where, w specify the total amount of energy absorbed by the specimen (in joules), t is the thickness of the specimen in mm and b_n is the sample rest width in mm at the notch base.

The samples revealed an impact strength value of the jute carpet/epoxy specimens in the present case. The higher impact primarily occurs because of strong reinforcement and matrix materials bonding. The carpet/epoxy composite has an impact strength of 6.15 kJ/m². The impact properties of the composite specimens also depend upon the type of carpet materials, configuration, fiber orientation, etc. Also, it is governed by the poor interface between reinforcement and epoxy, leading to severe fiber breakage and fiber pullout (Kamble et al. 2020; Sanjeevi et al. 2021).

Compression Test

The composite materials were cut into the ASTM D695 for the compression behavior during testing. To examine the mechanical performances, specimens of each composition with dimensions of 50.8 x 12.7 x 12.7 mm^3 were cut as per ASTM D695 standard. Composite compression tests were done using a METLOR material testing machine with a maximal capacity of 1000 kN to measure the strength of created materials. The sample's right edge was placed on the fixed jaw, and the moveable jaw was pressed downward.

At room temperature, the mechanical characteristics of epoxy/carpet waste composites were examined for compression behavior. Standard has been given the dry conditions data on the composites. It reveals the findings of the polymer samples developed from discarded carpet materials is 5 kN. Mechanical characteristics can be moderated due to the creation of cracks, fracture propagation, matrix debonding and enlargement of the plastic cavity, matrix deformation, and matrix pullout (Leopold et al. 2019; Foti 2013; Awal and Mohammadhosseini 2016). Previous investigations have shown that the compressive strength of carpet fiber/epoxy composites reduces after a 1200 exposure period (hours) (Afshar, Mihut, and Chen 2019). The enhanced compression characteristics required for feasible component preparation could be modified by the interfacial strength of epoxy material (Panchagnula and Kuppan 2019; Mechin, Keryvin, and Grandidier 2020; Alamri and Low 2012).

Superior epoxy-to-reinforcement interaction and increased load-bearing capacity can be acquired for the enhanced mechanical characteristics in developed samples. Scanning electron microscopy is necessary for examining the fragmented samples' surfaces. Figure 3.10 shows the SEM study of the cracked surface of the specimen. Carpet fiber and epoxy debonding are well shown in the sample, with their textures, porosity, and unevenness clearly remarked. Because of the stress, fiber breakage increased as a result of withdrawal, limiting the frying of the fibers (Sadrolodabaee, Claramunt, and Ardanuy 2021).

The waste carpet-based polymer composite is subjected to compression as part of a morphological study. Different failures were spotted in the scans since they were shown at 1 mm resolution. Figure 3.11 shows the epoxy/carpet

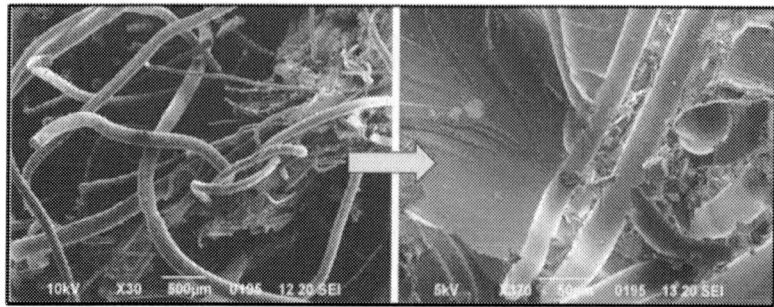

Figure 3.10. SEM analysis of fractured specimen (J. Kumar, Kumar, Kumar, et al. 2022).

Figure 3.11. Microscopic images (J. Kumar, Kumar, Jaiswal, et al. 2022).

composite created during the compression process. Different types of defects have been found (cavity, debonding, fiber pull out). Fractures, holes, and cracks are formed as fibers are pulled out. Compressive strength may be increased by using carpet fibers that may partially transmit stress across a fracture (Mastali and Dalvand 2016). The matrix fractured as a result of the strong pressure imposed by the fibers. Figure 3.11 shows the formation of many microcracks at the fiber matrix-matrix contact. Previous research on the impact of various defects caused during compression testing supported the present result (Mastali, Dalvand, and Sattarifard 2016; Awal and Mohammadhosseini 2016). It was also accompanied by microscopic images that proved the impact of flaws on compression characteristics. The mechanical characteristics of several carpet waste polymer composites are summarised in Table 3.2.

Table 3.2. Mechanical properties of different carpet waste polymer composites

Matrix	Reinforcement	Method	Mechanical Properties (MPa)			Remark	Ref.
			TS	FS	IS		
Epoxy	Nylon and Olefin waste carpet	VARTM	-	34-24	-	Nylon face fiber-based composite exhibits higher flexural strength than olefin	(Jain et al. 2012)
Epoxy and Polyester	Carpet waste jute yarn	Compression molding technique	-	111.7	40.2 KJ/m^2	It was observed that untreated water absorption composite shows the highest flexural and impact strength.	(Karaduman and Onal 2011)
Epoxy resin	Polyester/jute	Compression molding technique	161.78	-	1295.46 J/m	The results show that the polyester/jute hybrid composite obtained the higher mechanical strength	(Laranjeira et al. 2006)
Balsawood, PET foam, PP honeycomb	Jute/ Polypropylene	Compression molding technique and adhesive-free fabrication method	28.94	51.83	-	The study revealed that reinforcement of jute fiber could enhance the sandwich composite	(Karaduman and Onal 2016)
Epoxy resin	Flax/jute	Compression molding technique similar to hand layup	36.38	112.25	36.78 KJ/m^2	The authors stated that these composites could have high compatibility in roof sheets, automobiles, etc.	(Karthi et al. 2021)

Summary

The characterization of recycled waste carpets is the foundation of the current part, as was said at the outset, in order to create a structural composite material. Due to rising consumer demand, the carpet industry is growing every day, which also causes an increase in waste carpet production. The use of carpets and rugs has been expanded to include a variety of new functions, such as floor covering in homes, hospitals, airports, and commercial facilities. The task of managing carpet waste is the greatest challenge because carpets typically last 4-6 years. After that period, it began to degrade and became solid waste. Most carpet users dispose of their carpets in the form of contaminated landfills or garbage sites. Improper disposal can lead to many health and environmental hazards. Carpet waste is commonly dumped in landfills and rivers, where it directly harms humans and animals. In addition, it causes environmental pollution after incineration or landfill, as well as offensive odors in the environment.

In this chapter, the developed composites are mechanically characterized by performing tensile, flexural, compressive, and impact strength studies. An overall conclusion based on the knowledge gained can be drawn as follows.

- The mechanical properties of carpet-based composite considering four different materials: wool, polypropylene, nylon, and polyester were examined. Two samples of each material with different configurations were fabricated: FBBF and BFFB. Flexural and tensile tests were used to characterize the mechanical properties of the fabricated carpet waste composite samples.
- Wool material composite has a higher tensile strength of both laminate FBBF (18.2 MPa) and BFFB (16.4 MPa) among the developed composites.
- The flexural strength of the wool BFFB (34.40 MPa) sample is higher than FBBF (23 MPa) sample.
- The polypropylene FBBF (30.64 MPa) sample has lower flexural strength than BFFB (34 MPa)
- The flexural strength of the nylon FBBF (33.19 MPa) sample is higher than the BFFB (25.47 MPa) sample.
- The flexural strength of the polyester FBBF (28.23 MPa) sample is higher than that of the BFFB (26.61 MPa).

The results demonstrate a cost-effective and viable solution to carpet waste disposal. The resulting composites can be used to develop structural composites. The results of the current work demonstrate viable solutions for lightweight structural and decorative materials. The waste generated in the carpet sector is large and harmful to the environment. Limited data is available on the reuse of carpet waste. The results of current work may lead to new solutions for waste from the carpet and textile industry. The proposed work is a low-cost, eco-friendly solution for end-use applications.

References

Adak, Nitai Chandra, Suman Chhetri, Naresh Chandra Murmu, Pranab Samanta, Tapas Kuila, and Joong Hee Lee. 2018. "Experimental and Numerical Investigation on the Mechanical Characteristics of Polyethylenimine Functionalized Graphene Oxide Incorporated Woven Carbon Fibre/Epoxy Composites." *Composites Part B: Engineering* 156: 240–51. doi:10.1016/j.compositesb.2018.08.086.

Afshar, Arash, Dorina Mihut, and Pengyu Chen. 2019. "Effects of Environmental Exposures on Carbon Fiber Epoxy Composites Protected by Metallic Thin Films." *Journal of Composite Materials* 54 (2): 167–77. doi:10.1177/0021998319859051.

Alamri, H., and I. M. Low. 2012. "Effect of Water Absorption on the Mechanical Properties of Nano-Filler Reinforced Epoxy Nanocomposites." *Materials and Design* 42: 214–22. doi:10.1016/j.matdes.2012.05.060.

Alhuthali, A., I. M. Low, and C. Dong. 2012. "Characterisation of the Water Absorption, Mechanical and Thermal Properties of Recycled Cellulose Fibre Reinforced Vinyl-Ester Eco-Nanocomposites." *Composites Part B: Engineering* 43 (7): 2772–81. doi:10.1016/j.compositesb.2012.04.038.

Alrshoudi, Fahed, Hossein Mohammadhosseini, Mahmood Md Tahir, Rayed Alyousef, Hussam Alghamdi, Yousef R. Alharbi, and Abdulaziz Alsaif. 2020. "Sustainable Use of Waste Polypropylene Fibers and Palm Oil Fuel Ash in the Production of Novel Prepacked Aggregate Fiber-Reinforced Concrete." *Sustainability* 12 (12): 1–14. doi:10.3390/su12124871.

Awal, A. S. M. Abdul M. Abdul, and Hossein Mohammadhosseini. 2016. "Green Concrete Production Incorporating Waste Carpet Fiber and Palm Oil Fuel Ash." *Journal of Cleaner Production* 137: 157–66. doi:10.1016/j.jclepro.2016.06.162.

Burmistrov, I., N. Gorshkov, I. Ilinykh, D. Muratov, E. Kolesnikov, S. Anshin, I. Mazov, J. P. Issi, and D. Kusnezov. 2016. "Improvement of Carbon Black Based Polymer Composite Electrical Conductivity with Additions of MWCNT." *Composites Science and Technology* 129: 79–85. doi:10.1016/j.compscitech.2016.03.032.

Chaudhery, Sukanchan Palit, and Mustansar Hussain. 2019. "Nanomaterials for Environmental Science: A Recent and Future Perspective." In *Nanotechnology in Environmental Science*, edited by Chaudhery Mustansar Hussain and Ajay Kumar

Mishra, 1–872. Detection Science. Weinheim, Germany: John Wiley & Sons. doi:10.1002/9783527808854.ch1.

Chen, Dongdong, Guangyong Sun, Maozhou Meng, Xihong Jin, and Qing Li. 2019. "Flexural Performance and Cost Efficiency of Carbon/Basalt/Glass Hybrid FRP Composite Laminates." *Thin-Walled Structures* 142: 516–31. doi:10.1016/j.tws.2019.03.056.

Choobbasti, Asskar Janalizadeh, Mostafa Amozadeh Samakoosh, and Saman Soleimani Kutanaei. 2019. "Mechanical Properties Soil Stabilized with Nano Calcium Carbonate and Reinforced with Carpet Waste Fibers." *Construction and Building Materials* 211: 1094–1104. doi:10.1016/j.conbuildmat.2019.03.306.

Choudhary, Veena, and Anju Gupta. 2001. "Polymer/Carbon Nanotube Nanocompo-sites." In *Carbon Nanotubes - Polymer Nanocomposites*, edited by Siva Yellampalli, 65–90. Intech Open. doi:10.5772/18423.

Coleman, Jonathan N., Umar Khan, Werner J. Blau, and Yurii K. Gun'ko. 2006. "Small but Strong: A Review of the Mechanical Properties of Carbon Nanotube-Polymer Composites." *Carbon* 44 (9): 1624–52. doi:10.1016/j.carbon.2006.02.038.

Domun, N., H. Hadavinia, T. Zhang, T. Sainsbury, G. H. Liaghat, and S. Vahid. 2015. "Improving the Fracture Toughness and the Strength of Epoxy Using Nanomaterials- a Review of the Current Status." *Nanoscale* 7 (23): 10294–329. doi:10.1039/C5NR01354B.

Fan, Hailong, Lili Wang, Keke Zhao, Nan Li, Zujin Shi, Zigang Ge, and Zhaoxia Jin. 2010. "Fabrication, Mechanical Properties, and Biocompatibility of Graphene-Reinforced Chitosan Composites." *Biomacromolecules* 11 (9): 2345–51. doi:10.1021/bm100470q.

Foti, Dora. 2013. "Use of Recycled Waste Pet Bottles Fibers for the Reinforcement of Concrete." *Composite Structures* 96: 396–404. doi:10.1016/j.compstruct.2012.09.019.

Gang, Du. 2018. "The Effect of Surface Treatment of CF and Graphene Oxide on the Mechanical Properties of PI Composite." *Journal of Thermoplastic Composite Materials* 31 (9): 1219–31. doi:10.1177/0892705717734606.

Ghorai, Suman. 2013. "Chemical, Physical and Mechanical Properties of Nanomaterials and Its Applications." *University of Iowa*. University of Iowa. doi:10.17077/etd.5r1xf0y8.

Islam, Shafiqul, and Gajanan Bhat. 2019. "Environmentally-Friendly Thermal and Acoustic Insulation Materials from Recycled Textiles." *Journal of Environmental Management* 251: 109536. doi:10.1016/j.jenvman.2019.109536.

Jain, Abhishek, Gajendra Pandey, Abhishek K. Singh, Vasudevan Rajagopalan, Ranji Vaidyanathan, and Raman P. Singh. 2012. "Fabrication of Structural Composites from Waste Carpet." *Advances in Polymer Technology* 31 (4): 380–89. doi:10.1002/adv.20261.

Jiménez-Pérez, Víctor Manuel, Oxana V. Kharissova, and Blanca M. Muñoz Flores. 2017. "Environmental Applications of Iron-Containing Nanomaterials: Synthetic Routes, Structures, Compositions and Properties." In *Advanced Environmental Analysis: Applications of Nanomaterials*, edited by Chaudhery Mustansar Hussain and Boris

Kharisov, 1:193–220. Detection Science. Cambridge, United States: The Royal Society of Chemistry.

Kamble, Zunjarrao, Bijoya Kumar Behera, Teruo Kimura, and Ino Haruhiro. 2020. "Development and Characterization of Thermoset Nanocomposites Reinforced with Cotton Fibres Recovered from Textile Waste." *Journal of Industrial Textiles* 51 (2): 2026S-2052S. doi:10.1177/1528083720913535.

Karaduman, Y., and L. Onal. 2011. "Water Absorption Behavior of Carpet Waste Jute-Reinforced Polymer Composites." *Journal of Composite Materials* 45 (15): 1559–71. doi:10.1177/0021998310385021.

Karaduman, Y. and, and L. Onal. 2016. "Flexural Behavior of Commingled Jute/Polypropylene Nonwoven Fabric Reinforced Sandwich Composites." *Composites Part B: Engineering* 93: 12–25. doi:10.1016/j.compositesb.2016.02.055.

Karthi, N., K. Kumaresan, S. Sathish, L. Prabhu, S. Gokulkumar, D. Balaji, N. Vigneshkumar, et al. 2021. "Effect of Weight Fraction on the Mechanical Properties of Flax and Jute Fibers Reinforced Epoxy Hybrid Composites." *Materials Today: Proceedings* 45 (9): 8006–10. doi:10.1016/j.matpr.2020.12.1060.

Kumar, Amit, Kamal Sharma, and Amit Rai Dixit. 2019. "A Review of the Mechanical and Thermal Properties of Graphene and Its Hybrid Polymer Nanocomposites for Structural Applications." *Journal of Materials Science* 54 (8): 5992–6026. doi:10.1007/s10853-018-03244-3.

Kumar, Jogendra, Kaushlendra Kuldeep Kaushlendra Kumar, Balram Jaiswal, Kaushlendra Kuldeep Kaushlendra Kumar, and Rajesh Kumar Verma. 2022. "Investigation on the Physio-Mechanical Properties of Carpet Waste Polymer Composites Incorporated with Multi-Wall Carbon Nanotube (MWCNT)." *Journal of the Textile Institute*, 1–10. doi:10.1080/00405000.2022.2062860.

Kumar, Jogendra, Kuldeep Kaushlendra Kumar, Kuldeep Kaushlendra Kumar, Balram Jaiswal, and Rajesh K. Verma. 2022. "Development of Waste Carpet (Jute) and Multi-Wall Carbon Nanotube Incorporated Epoxy Composites for Lightweight Applications." *Progress in Rubber Plastics and Recycling Technology* 38 (3): 247–63. doi:10.1177/14777606221110252.

Kumar, M. S.Senthil, N. Mohana Sundara Raju, P. S. Sampath, and L. S. Jayakumari. 2014. "Effects of Nanomaterials on Polymer Composites - An Expatiate View." *Reviews on Advanced Materials Science* 38 (1): 40–54.

Kumar Verma, Rajesh, Balram Jaiswal, Rahul Vishwakarma, Kuldeep Kaushlendra Kumar, Kuldeep Kaushlendra Kumar, Rajesh Kumar Verma, Balram Jaiswal, Rahul Vishwakarma, and Kuldeep Kaushlendra Kumar. 2021. "Polymer Composite Developed from Discarded Carpet for Light Weight Structural Applications: Development and Mechanical Analysis." *E3S Web of Conferences* 309: 01154 (1-5). doi:10.1051/e3sconf/202130901154.

Laranjeira, E., L. H. De Carvalho, S. M.D.L. Silva, and J. R.M. D'Almeida. 2006. "Influence of Fiber Orientation on the Mechanical Properties of Polyester/Jute Composites." *Journal of Reinforced Plastics and Composites* 25 (12): 1269–78. doi:10.1177%2F0731684406060577.

Leopold, Christian, Gordon Just, Ilja Koch, Andreas Schetle, Julia B. Kosmann, Maik Gude, and Bodo Fiedler. 2019. "Damage Mechanisms of Tailored Few-Layer

Graphene Modified CFRP Cross-Ply Laminates." *Composites Part A: Applied Science and Manufacturing* 117: 332–44. doi:10.1016/j.compositesa.2018.12.005.

Mastali, M., and A. Dalvand. 2016. "The Impact Resistance and Mechanical Properties of Self-Compacting Concrete Reinforced with Recycled CFRP Pieces." *Composites Part B* 92: 360–76. doi:10.1016/j.compositesb.2016.01.046.

Mastali, M., A. Dalvand, and A. R. Sattarifard. 2016. "The Impact Resistance and Mechanical Properties of Reinforced Self- Compacting Concrete with Recycled Glass Fi Bre Reinforced Polymers." *Journal of Cleaner Production* 124: 312–24. doi:10.1016/j.jclepro.2016.02.148.

Mechin, P.-Y., V. Keryvin, and Jean Claude Grandidier. 2020. "Limitations on Adding Nano-Fillers to Increase the Compressive Strength of Continous Fibre/Epoxy Matrix Composites." *Composites Science and Technology* 192: 108099. doi:10.1016/j.compscitech.2020.108099.

Mihut, Corina, Dinyar K. Captain, Francis Gadala-Maria, and Michael D. Amiridis. 2001. "Review: Recycling of Nylon from Carpet Waste." *Polymer Engineering and Science* 41 (9): 1457–70. doi:10.1002/pen.10845.

Miraftab, Mohsen, Richard Horrocks, and Colin Woods. 1999. "Carpet Waste, an Expensive Luxury We Must Do Without." *Autex Research Journal* 1 (1): 1–7. doi:10.1533/9780857092991.3.173.

Mishra, Kunal, and Ranji Vaidyanathan. 2019a. "The Influence of Nanoclay on the Flame Retardancy and Mechanical Performance of Recycled Carpet Composites." *Recycling* 4 (2): 22 (1-10). doi:10.3390/recycling4020022.

Mishra, Kunal, and Ranji K. Vaidyanathan. 2019b. "Application of Recycled Carpet Composite as a Potential Noise Barrier in Infrastructure Applications." *Recycling* 4 (1): 9 (1-11). doi:10.3390/recycling4010009.

Mohammadhosseini, Hossein, A. S. M. Abdul Awal, and Jamaludin B. Mohd. 2017. "The Impact Resistance and Mechanical Properties of Concrete Reinforced with Waste Polypropylene Carpet Fibres." *Construction and Building Materials* 143: 147–57. doi:10.1016/j.conbuildmat.2017.03.109.

Mohammadhosseini, Hossein, and Jamaludin Mohamad Yatim. 2017. "Microstructure and Residual Properties of Green Concrete Composites Incorporating Waste Carpet Fibers and Palm Oil Fuel Ash at Elevated Temperatures." *Journal of Cleaner Production* 144: 8–21. doi:10.1016/j.jclepro.2016.12.168.

Naebe, Minoo, Jing Wang, Abbas Amini, Hamid Khayyam, Nishar Hameed, Lu Hua Li, Ying Chen, and Bronwyn Fox. 2014. "Mechanical Property and Structure of Covalent Functionalised Graphene/Epoxy Nanocomposites." *Scientific Reports* 4: 4375 (1-7). doi:10.1038/srep04375.

Ning, Huiming, Jinhua Li, Ning Hu, Cheng Yan, Yaolu Liu, Liangke Wu, Feng Liu, and Jianyu Zhang. 2015. "Interlaminar Mechanical Properties of Carbon Fiber Reinforced Plastic Laminates Modified with Graphene Oxide Interleaf." *Carbon* 91: 224–33. doi:10.1016/j.carbon.2015.04.054.

Onal, L., and Y. Karaduman. 2009. "Mechanical Characterization of Carpet Waste Natural Fiber-Reinforced Polymer Composites." *Journal of Composite Materials* 43 (16): 1751–68. doi:10.1177%2F0021998309339635.

Pal, Hemant, Vimal Sharma, and Manjula Sharma. 2014. "Influence of Functionalization on Mechanical and Electrical Properties of Carbon Nanotube-Based Silver Composites." *Philosophical Magazine* 94 (13): 1478–92. doi:10.1080/14786435.2014.892221.

Panchagnula, Kishore Kumar, and Palaniyandi Kuppan. 2019. "Improvement in the Mechanical Properties of Neat GFRPs with Multi-Walled CNTs." *Journal of Materials Research and Technology* 8 (1): 366–76. doi:10.1016/j.jmrt.2018.02.009.

Pathak, Abhishek K., Munu Borah, Ashish Gupta, T. Yokozeki, and Sanjay R. Dhakate. 2016. "Improved Mechanical Properties of Carbon Fiber/Graphene Oxide-Epoxy Hybrid Composites." *Composites Science and Technology* 135: 28–38. doi:10.1016/j.compscitech.2016.09.007.

Phiri, Josphat, Patrick Gane, and Thad C. Maloney. 2017. "General Overview of Graphene: Production, Properties and Application in Polymer Composites." *Materials Science and Engineering B* 215: 9–28. doi:10.1016/j.mseb.2016.10.004.

Rafiee, Mohammad, Fred Nitzsche, Jeremy Laliberte, Jules Thibault, and Michel R. Labrosse. 2019. "Simultaneous Reinforcement of Matrix and Fibers for Enhancement of Mechanical Properties of Graphene-Modified Laminated Composites." *Polymer Composites* 40 (S2): E1732–45. doi:10.1002/pc.25137.

Reis, João Marciano Laredo dos. 2009. "Effect of Textile Waste on the Mechanical Properties of Polymer Concrete." *Materials Research* 12 (1): 63–67. doi:10.1590/S1516-14392009000100007.

Sadrolodabaee, Payam, Josep Claramunt, and M. Ardanuy. 2021. "A Textile Waste Fiber-Reinforced Cement Composite: Comparison between Short Random Fiber and Textile Reinforcement." *Materials* 14 (13): 3742 (1-17). doi:10.3390/ma14133742.

Sanjeevi, Sekar, Vigneshwaran Shanmugam, Suresh Kumar, Velmurugan Ganesan, Gabriel Sas, Deepak Joel Johnson, Manojkumar Shanmugam, et al. 2021. "Effects of Water Absorption on the Mechanical Properties of Hybrid Natural Fibre/Phenol Formaldehyde Composites." *Scientific Reports* 11: 13385. doi:10.1038/s41598-021-92457-9.

Stewart, Richard. 2009. "Lightweighting the Automotive Market." *Reinforced Plastics* 53 (3): 14–21. doi:10.1016/S0034-3617(09)70078-5.

SukanchanPalit, and Chaudhery MustansarHussain. 2018. *Handbook of Nanomaterials for Industrial Applications*. Edited by Chaudhery Mustansar Hussain. Micro and Nano Technologies. Amsterdam, Netherlands: Elsevier Ltd. doi:10.1016/C2016-0-04427-3.

Teli, M. D. 2018. "Finishing of Carpets for Value Addition." In *Advances in Carpet Manufacture*, edited by K. K. Goswami, Second Edi, 175–211. United Kingdom: Elsevier Ltd. doi:10.1016/B978-0-08-101131-7.00010-1.

Wang, Youjiang, Mehmet Ucar, and Youjiang Wang. 2014. "Utilization of Recycled Post Consumer Carpet Waste Fibers as Reinforcement in Lightweight Cementitious Composites." *International Journal of Clothing Science and Technology* 23 (4): 24–248. doi:10.1108/09556221111136502.

Yalcin-enis, Ipek, Merve Kucukali-ozturk, and Hande Sezgin. 2019. "Risks and Management of Textile Waste." In *Nanoscience and Biotechnology for Environmental Applications*, edited by K. M. Gothandam, Shivendu Ranjan, Nandita Dasgupta, and

Eric Lichtfouse, 29–53. Switzerland: Springer Nature Switzerland. doi:10.1007/978-3-319-97922-9.

Yuan, Xiaomin, Bo Zhu, Xun Cai, Kun Qiao, Shengyao Zhao, and Min Zhang. 2019. "Nanoscale Toughening of Carbon Fiber-Reinforced Epoxy Composites through Different Surface Treatments." *Polymer Engineering and Science* 59 (3): 625–32. doi:10.1002/pen.24978.

Zhang, Yajie, Dan Deng, Kun Lu, Jianqi Zhang, Benzheng Xia, Yifan Zhao, Jin Fang, and Zhixiang Wei. 2015. "Synergistic Effect of Polymer and Small Molecules for High-Performance Ternary Organic Solar Cells." *Advanced Materials* 27 (6): 1071–76. doi:10.1002/adma.201404902.

Chapter 4

Physical and Morphology Properties of Composites Developed from Carpet Waste

This chapter focuses on the physical and morphological performance of the polymer composites developed from discarded carpets. These performances play an important role in identifying the realizable properties and causes of failures. The end uses of these discarded carpets can support the development of components exposed to humidity conditions, flame retardancy issues, internal defect generation, etc. This information highlights the importance and research-oriented aspects of the proposed book. Communicate the necessary knowledge about the composite properties of recycled waste carpets and their scientific implementation. In this chapter, the structure of carpet waste/epoxy composites analysis was described. Based on past research studies and achievements of waste carpet composite materials, physical properties and morphological characteristics (water absorption, flame retardancy, temperature) are investigated.

Morphological Characterization

The microstructure and morphology of substrates, particularly at the surface, are usually required to be well understood in high-value production since they influence safe loading conditions and longevity (Hajsman et al. 2019). Morphological characterization of the developed composite has been performed to characterize based on the thermal properties of the material and microstructural analysis (Tohidi et al. 2018; Wang et al. 2020). This chapter investigates the chemical composition, phase, and crystallinity of the discarded composite samples.

X-Ray Diffraction (XRD) Analysis

The X-ray diffraction analysis, often known as XRD, is a microstructural and nondestructive testing procedure and is an advanced technique for characterizing the structure, phase, crystal direction, and other structural parameters, such as average kernel sizes, crystallinity, strain, and crystal defects (Verma et al. 2022; Mishra and Vaidyanathan 2019; Kumar and Verma 2021). It is based on the constructive interference of monochromatic X-rays with a crystallized sample and follows the Bragg's law. Decelerating electrically charged particles with sufficient energy produce X-rays, which are shorter wavelength electromagnetic radiation. Diffraction is the result of the interaction of X-rays with an object, which is detected, processed, and tallied in a process known as electron tomography in XRD. When diffracted light is scattered at different angles, it creates a charted diffraction pattern. XRD peaks are created by a monochromatic X-rays beam dispersed off each set of lattice planes at specified angles. The atomic locations dictate the maximum intensity of the lattice planes. Fully automated optical alignment under computer control Ultima-IV machines shown in Figure 4.1 was used to X-ray diffract the developed carpet waste structural composites.

Figure 4.1. XRD analyzer.

XRD assessment was conducted for epoxy-based waste carpet composite of different materials and illustrated. For the polyester material in Figure 4.2

(a), remarked spectra confined three peaks it has primarily recognized for the polyester reinforced with epoxy. The ordinate describes the XRD graph with extreme refractile intensity, and the abscissa specifies the diffraction angle. It has acknowledged the presence of the polyester reinforce epoxy with its developed peaks at $2\theta = 24°$ and the small peaks at $28°$ and $43°$. In polypropylene (PP) polymer composite, the XRD analysis is mostly preferred to compute the degree of epoxy and PP reinforcement. The existence of high peaks for PP epoxy composites is at higher peak $2\theta=19.5°$ and lower peaks $2\theta=17°$ and $44°$. Figure 4.2 (b) confirms the presence of polypropylene in epoxy (Kumar, Kumar, Kumar, et al. 2022; Verma et al. 2022).

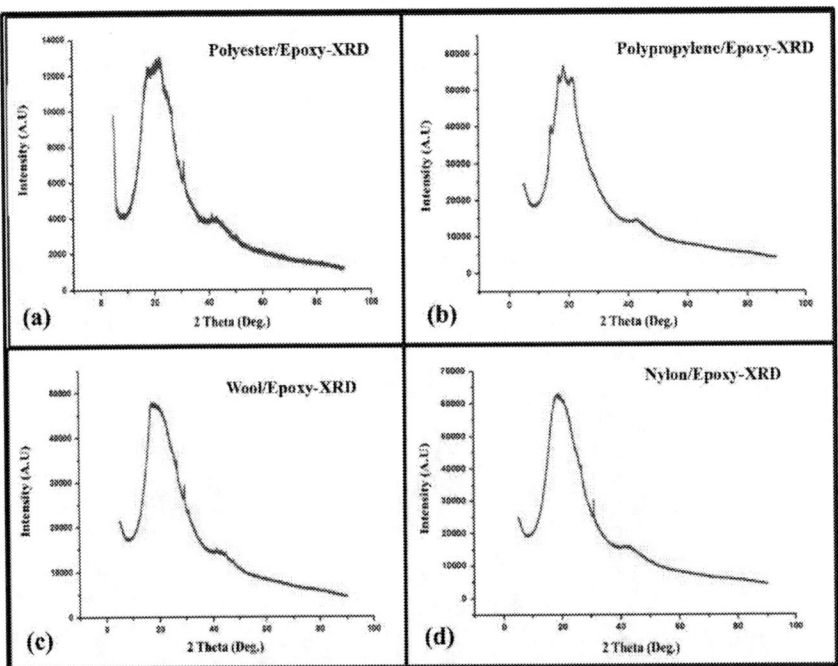

Figure 4.2. XRD analysis of epoxy-based carpet waste composite (a) Polyester (b) Polypropylene (c) Wool (d) Nylon.

The XRD output of wool reinforcement in epoxy is shown in Figure 4.2 (c). The wool and epoxy peak are visible in the fabricated composite of wool reinforced with epoxy, as shown in Figure 4.2 (c). According to the figure, a solid solution was produced by mixing wool and epoxy, ensuing in the development of wool reinforcement in epoxy, and finally, the composite's peak can be seen. The crystalline features of nylon-modified polymer (epoxy) composites are appropriately shown in the diffractogram at 2θ =30°, as shown in Figure 4.2 (d). In the X-ray pattern, wool reinforcement in epoxy can be seen at 2θ = 25.1° peak. This demonstrates the presence of wool reinforcement in epoxy composite particles in the fabricated composite.

Fourier Transform Infrared (FTIR) Analysis

The Fourier Transform Infrared Analysis is a convenient analytical tool used for recognizing functional groups and covalent bonding information. FTIR spectrometer shown in Figure 4.3 was used for spectroscopy analysis and to identify chemical compounds of the developed carpet waste structural composites. FTIR explores the quantitative and qualitative findings for organic and inorganic materials. By generating an infrared light absorption spectrum, it computes the existing chemical bonds in a molecule. The spectra contain a profile, an exclusive chemical impression that will be used to screen and scan specimens.

Figure 4.3. Spectrum-2 FTIR spectrometer.

FTIR spectra of polymer (epoxy) composite modified by discarded carpets are shown in Figure 4.4 (a). The assignments and wavenumbers of FTIR transmissivity bands were summed up in Table 4.1. At a wave number of 3340 cm^{-1}, the existence of the ≡C-H groups was detected in the stretching vibrations. The bonding of hydrogen between polyester samples is more likely to happen from this standpoint, as mentioned above. The C-H group weak stretching vibrations such as -CH$_2$ and -CH$_3$ are described by the transmissivity bands in the range 2900–3100 cm^{-1}. After curing the polyester with epoxy, the transmissivity bands of the stretching vibrations of the C-H groups had virtually the same transmissivity bands. Transmittivity was examined at 1716 cm^{-1} during the infrared spectrum investigation due to the significant stretched vibrations of the C=O aldehyde group. The spectra of carpet (polyester)/epoxy-based composites exhibit subtle changes, resulting in a shift towards higher frequencies. Weak bands found in the cured polyester spectrum at 1508 cm^{-1} can be assigned to the aromatic ring C= C. The strong band at 1243 cm^{-1} that appears in the polyester spectrum is due to C-O-C group vibrations. From the graph, it can be concluded that the C-OH group at 1034 cm^{-1} has good stretch interactions. One intensive band at 752 cm^{-1} was attributed to C-Cl strong stretching vibrations in the polyester/epoxy composite spectrum. After 752 cm^{-1} wavenumbers, halogen groups are observed to appear continuously until wavenumbers of 500 cm^{-1}.

Table 4.1. Assignment of main infrared transmissivity of wave numbers

Wave Numbers (cm^{-1})	Assignment
3340	≡C-H Strong Stretch
2925	-C-H weak
2874	-C-H aldehyde Variable
1716	C=O aldehyde Strong
1508	C=C aromatic weak
1243	C-O-C stretch strong
1034	C-OH stretch strong
752	C-Cl strong
500	C-I strong

Figure 4.4 (b) shows the absence of peaks in a single bond area (1700-4000 cm^{-1}). While no broad absorption band was noticed. Also, notifying there is no hydrogen bond in the material. There is some very small absorption that can be shown in the range of 500-1500 cm^{-1}. Small peaks were identified at 1626 cm^{-1} medium stretching of C=C, 1380 cm^{-1} medium stretching of C-H, 1125 cm^{-1} strong stretching of C-O, and 870 cm^{-1} bending of C-H. A sharp bond at about 500 cm-1 indicates the stretching of the C-I halogen compound.

In Figure 4.4 (c), for the wool composite higher wavenumber region, O-H stretching vibration band occurred in the range 3200-3550 cm^{-1}. At 3000 cm^{-1}, weak O-H stretching occurs. It can be seen from the spectra of epoxy-based carpet (wool) waste the stretching vibrations of the N-O group of the epoxy resin at 1500 cm^{-1}. In the spectra, the band at 1250 cm^{-1} corresponds to the stretching of aromatic ester C-O. Similarly, a prominent band at 1085 cm^{-1} corresponds to the aliphatic ether C-O. The band at 500 cm^{-1} indicates the stretching of the C-I halogen compound. At 750 cm^{-1}, strong C-H bending occurs. A sudden sharpness in the peak at the 500 cm^{-1} region reveals a stretching of the C-I halogen complex.

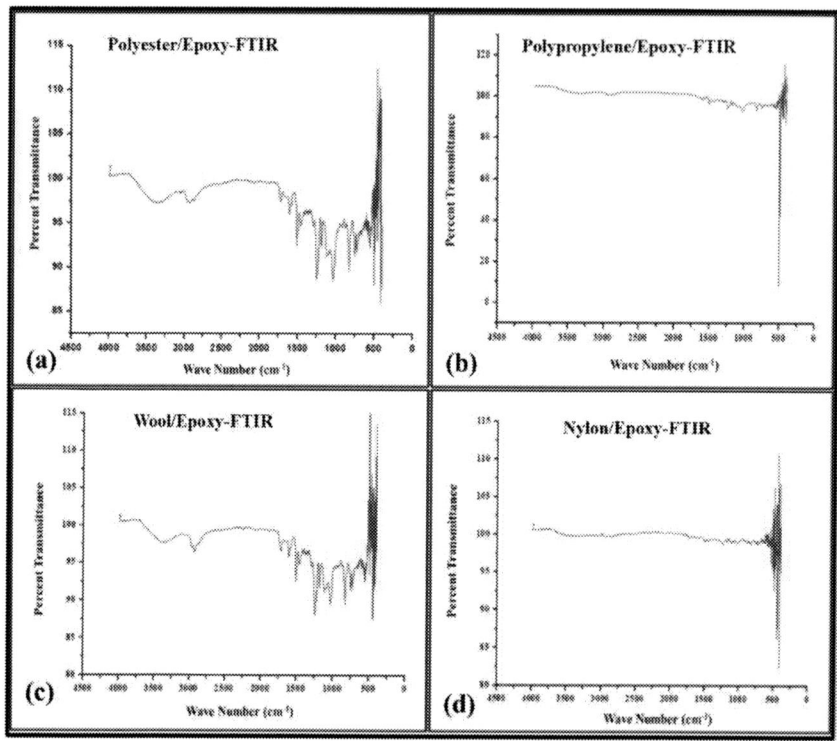

Figure 4.4. FTIR analysis of epoxy-based carpet waste composite (a) polyester (b) polypropylene (c) wool (d) nylon.

Whereas, for the nylon composite, shown in Figure 4.4 (d), there are no peaks in a single bond area (1500-4000 cm^{-1}). The lack of any broad absorption band illustrates that the developed Nylon based composite material is bereft of hydrogen bonds. The band range of 500-600 cm^{-1} is related to the halogen compound that is stretching of C-I. According to the data tabulated in Table 4.2, it has also been confirmed in other research.

Table 4.2. Morphology properties developed from different carpet waste polymer composites

Sr. No.	Types of carpet waste	Matrix	XRD	Ref.
1.	Nylon 66-based cut loop type of carpet waste	Mixture of Epoxy + Polyester + Vinyl ester	$2\theta = 10°$ and $20°$ Represents matrix and macro reinforcement $2\theta = 20\text{-}80°$ sharp peaks represent OMMT clay powder	(Mishra and Vaidyanathan 2019)

Physical Properties Characterization

In prior studies based on carpet waste management, there has been limited research on the manufacturing and engineering properties analysis of thermoset (epoxy) composites made from discarded waste. Carpet waste incineration, damping, and fuel usage are the primary subjects of most studies. The engineering industry focuses on recovering or reusing the produced waste in the environment, society, and trade. A lot of garbage is generated by these industries, which are both large and solid. The traditional recycling approach is unsuitable for the carpet industry due to expense, non-biodegradability, and environmental concerns. As a result, the current study focused primarily on the re-utilization of discarded carpets to generate polymer composites. Reusing old jute carpets has been attempted to decrease trash caused by carpet application. Developing lightweight structural composites in sound barriers, roadside barriers, insulating boards, frames, and so on might be a realistic waste management solution to establish a sustainable environment.

These performances play an important role in identifying realizable properties and failures and causes of failures in an application. The end uses of these discarded carpets can support the development of components exposed to humidity conditions, flame retardancy issues, internal defect generation, etc. This information highlights the importance and research-oriented aspects of the proposed book. Communicate the necessary knowledge about the composite properties of recycled waste carpets and its scientific implementation. In this chapter, the structure of carpet waste/epoxy composites analysis was described. Based on past research studies and achievements of waste carpet composite materials, physical properties and morphological characteristics (water absorption, flame retardancy, temperature) are investigated. Therefore, a highly efficient and cost-effective solid waste management approach is needed in view of the generated waste. The production of lightweight polymer components through the reuse of waste carpets was observed as a feasible solution in the prior work (Pakravan and Memarian 2015; Jaiswal, Verma, and Mishra 2022; Aghaee and Foroughi 2013; Kumar Verma et al. 2021). This finding in the current book describes the fabrication of cost-effective and durable polymer composites in a vacuum environment using the resin transfer molding (RTM) method. This created vacuum environment is observed as an effective way to diffuse the epoxy in the carpet laminates without contaminants and impurities. Due to the combined qualities of both materials, the completed product can be used for sound absorption, road barriers, wall panels/tiles, etc. Using the VARTM method to manufacture the polymer composites from waste jute carpets, provides satisfactory results, which are described in the below sections. A pioneering method was proposed to study the outcome of discarded recycling of carpet waste materials in polymers.

Water Absorption Test

The prepared sample (composites) were cut according to ASTM D570 (79.2 x 25.4 x 12.7 mm^3) specimens and immersed in distilled water for 24 hours for around ten days at room temperature. Figure 4.5 (a) shows how each sample was dried with tissue at regular intervals (every 24 hours) before being weighed on an electronic scale. According to the following equation (4.1), water content (Mt) was calculated:

$$M_t\% = \frac{W_f - W_i}{W_i} \tag{4.1}$$

Here $M_t\%$ = % water content
W_f = Sample weight (final)
W_i = Sample weight (initial)

The produced composites were examined for water absorption in this experimental investigation. Figure 4.5 (b) shows the water-socked results of discarded composites. The data on the water content (Mt %) absorbed by each sample was collected over the course of ten days of testing. The water intake of all specimens rises gradually (up to 5-6 days). Initially, the water socking rate was high and linear, but after saturation was attained, it progressively slowed down. The composite absorbed a negligible amount of liquid during its immersion duration. With carpet waste reinforcement, the sample has enhanced surface absorption of moisture and improved adhesion between fiber and epoxy. The combination of fiber's hydrophobic nature and composite's compressed densification results in a composite with a low level of porosity. It was found that the composite had been altered due to the synergetic impact of the presence of carpet waste reinforcement (densified compact, hydrophilic nature). Polymer might be permitted barrier mechanisms with reduced water permeability into composite interfacial, improving its efficiency in suppressing water absorption because of the varied carpet materials' epoxy matrix contents. As a result, the absorbent has a larger water content than typical composites.

Swelling Analysis

As mandated by ASTM D570, the produced sample was assessed for swelling rate. While immersing the samples in distilled water at room temperature, they had to be measured and recorded. The samples were dried on tissue paper for 24 hours before being measured for thickness. Figure 4.5 (c) shows the outcomes of the swelling (thickness) test over ten days. Calculating the proportion of swelling $S_t\%$ was evaluated using equation (4.2):

$$S_t\% = \frac{T_f - T_i}{T_i} \qquad (4.2)$$

Here $S_t\%$ = % of thickness swelling
T_f = Final dimension (thickness)
T_i = Initial dimension (thickness)

As shown in Figure 4.5 (d), the swelling properties of the polymer composite have been enhanced using recovered carpet waste (Kumar, Kumar, Jaiswal, et al. 2022). The impact of thickness swelling on epoxy/carpet composites is examined during a 10-day target period. Composites made from carpet waste were investigated for their thickness-swelling properties. There was less equilibrium thickness swelling in the composites' carpet waste reinforcement. Figure 4.5 (d) shows that the larger the amount of water intake in the early stages, the greater the rate of thickness swelling. Swelling will return to normal after 4-5 days of immersion. The hydrophobic nature of the large portion of the reinforcement utilized in waste fiber/polymer composites ensures the stability of these composites. The epoxy materials' reinforcement serves as a swelling resistance (increased interfacial); therefore, a small swelling might result from this reinforcement (Tong, Sun, and Tan 2008; Ning et al. 2015; He et al. 2018; Zhang et al. 2016). The results of prior investigations have all pointed to the same conclusion.

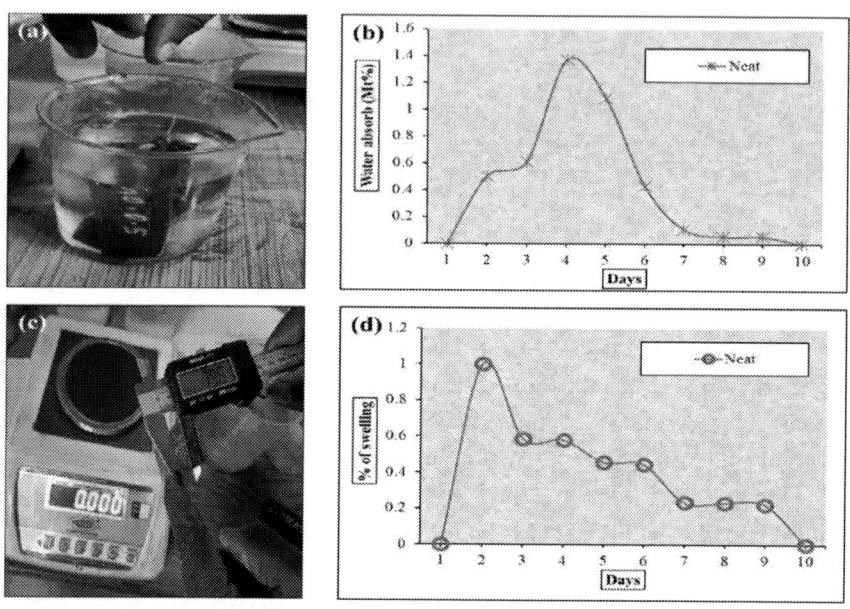

Figure 4.5. (a) Water absorption setup, (b) Water absorb content (Mt%), (c) Measurement of weight and swelling effect, and (d) Swelling % of the content (S_t%).

Polymer composites, as compared to pristine polymers, show less swelling because the length of the polymer chain between cross-links has been reduced. The greater surface area between the carpet (fiber) and matrix aids in water socking (capillary water uptake) (Javidparvar, Naderi, and Ramezanzadeh 2019). Improved interconnection of fiber and matrix reduces capillary water absorption by using epoxy modification. Fiber reinforcement has been proven in previous studies to enhance the interface. Karaduman et al. (Karaduman and Onal 2011) found that jute/matrix composites may be treated to prevent water absorption. The experiment revealed that brittleness increased water absorption with numerous pores because of the disorganized tendency for reinforcing (Awal and Mohammadhosseini 2016; Verma et al. 2022). Table 4.3 describes the various types of carpet waste and their respective mechanical properties.

Table 4.3. Physical properties Polymer composites developed from different carpet waste

Sr. No.	Types of carpet waste	Matrix	Absorption	Sample	Ref.
1.	Polypropylene carpet waste	Ordinary Portland Cement	2.5 mm	1.25% Wt. Sample	(Mohammadhosseini, Tahir, and Sayyed 2018)
		Palm Oil Fuel Ash	0.98 mm	1.25% Wt. Sample (20% POFA)	
2.	Jute carpet waste	Epoxy	~10% wt. gain	S10 W2 E-1	(Karaduman and Onal 2011)
		Polyester	~4.5% wt. gain	S10 W2 P-1	

Thermal Analysis

Relatively little research exists on the manufacturing and properties analysis (mechanical) of the carpet waste polymer composite, according to a prior study on carpet waste treatment. Most studies are focused on managing carpet waste, such as using carpet fuel, dampening, incineration, etc. In terms of waste management, the carpet and textile manufacturing sector must find a way to reuse or recycle the trash created. These industries produce enormous amounts of waste debris. The typical recycling approach is not ideal for the carpet industry because of its high cost, lack of biodegradability, and environmental concerns. As a result, the focus of this research was on the development of polymer composites using discarded carpets. Recycled jute carpet has been used to decrease the amount of trash created by carpet applications. For a more sustainable environment, developing lightweight structural composites for use like sound barriers, wall tiles, toys, and frames might be a helpful waste management strategy.

Thermal Gravimetric Analysis (TGA)

Figure 4.6 shows TGA 55 machine was used for the thermal analysis of the developed carpet waste structural composites. The Thermogravimetric analysis, often known as thermal gravimetric analysis (TGA), is a thermal analytical method that determines the mass of a sample as temperature changes over time. In particular, this study gives information on physical phenomena like the transitional phase, absorption, and desorption of substances.

Figure 4.6. Thermogravimetric analyzer.

Using a nitrogen environment, the thermal stability of the test sample was evaluated by comparing the weight loss with heat. It is shown in TGA that the epoxy-based waste carpet composite material has a high thermal conductivity. There are three main stages of reaction in deterioration, as shown in Figure 4.7. The loss of water or the breakdown of contaminants caused the first stage at 248°C. The weight reduction fraction was approximately 6.76%, as shown in Figure 4.7 (a). The noticeable peak is part of the second degradation process, which is connected to the breakdown of the polymeric portion. After this, the polymeric fraction started to degrade significantly, and there was significant heat degradation. The maximum weight loss occurred when the resin degraded, causing a temperature increase of roughly 109°C. At 600°C or thereabouts, the resins stopped degrading, losing 77% of their weight. Finally, it degrades to around 91.77% of its original weight at 800°C (Verma et al. 2022).

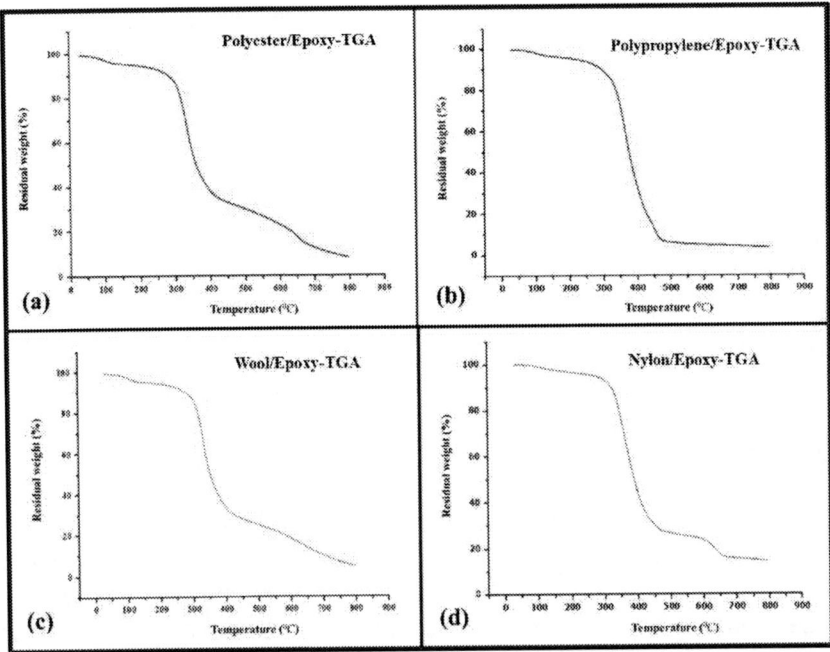

Figure 4.7. TGA analysis of epoxy-based carpet waste composite (a) polyester (b) polypropylene (c) wool (d) nylon.

Figure 4.7 (b) shows the TGA curves of the epoxy-based carpet (polypropylene) waste composite. Due to the evaporation of moisture and other volatiles at 100°C, a small weight loss was observed in the first step. The loss of weight fraction in the first step was around 10% (Karaduman and Onal 2011). The actual decay occurs in the second step at 300°C, caused by the thermal degradation of propylene. Following this stage, there was a massive amount of thermal degradation due to the degradation of the polymeric fraction. The decay of the resins ended at approximately 470°C, with 93.86% weight loss. Finally, at around 800°C, it degrades about 97.84% weight loss of the whole sample.

Figure 4.7 (c) displays the TGA curves of the polymer wool (waste) composite. During the first step, a slight weight loss was detected at the temperature of 100°C due to the evaporation of moisture and other volatiles (Verma et al. 2022). The loss of weight fraction in the first step was around 10%. The actual decay occurs in the second step at 290°C, caused by wool's thermal degradation. After this point, massive thermal degradation of polymeric material occurred. The deterioration of the resins ended at about 550°C, with 84% weight loss. Lastly, at about 800°C, it degrades nearly 96.84% weight loss.

Figure 4.7 (d) displays the TGA curves of the epoxy-based carpet (Nylon) waste composite. First, at 100°C, moisture and other volatiles evaporated, causing a little weight loss. The loss of weight fraction in the first step was around 8%. The actual decay occurs in the second step at 320°C, which is caused by the thermal degradation of Nylon. The temperature rose to 600°C, and the resin got decomposed. At 650°C, the resins began to deteriorate, losing 84% of their weight. The specimen declines by roughly 91.80% weight loss at about 800°C.

Fire Retardants

Combustible materials are prevented from igniting by using fire retardants, which slow down the burning process and do away with other potential sources of fire and conflagration. It is necessary to perform fire retardant experiments on the fabricated samples of polymers. It may be utilized for structural reasons to confirm the well-being of both human beings and goods. This test was performed as per the ASTM D635 standard for developed specimens. Thirty seconds of torch lighting is followed by 30 seconds of analyzing sample self-burning in the horizontal fire arrangement (Kumar, Kumar, Jaiswal, et al. 2022). Figure 4.8 depicts the setup used to study fire retardants.

The addition of carpet waste reinforcement delayed the initial drop in horizontal burning. Reinforcement has an essential role in increasing the fire-retardant qualities of carpet waste composites, as shown by the findings of the test samples. The synergistic effects can be explained by the physical effects caused by the particle arrangement on the composite's surface during polymer burning and ablation.

The high surface area of discarded carpet fiber is likely responsible for these fibers' catalytic effects. When there are more particles, there will be more places where the polymer chain might get limited; also, detaching the polymer chain will become highly complicated and demand a higher level of heat energy for pyrolysis. As a result of the polymer network's depth, the extra reinforcements and the polymer chain have a superior interfacial connection since the polymer chain has very little mobility (Pathak et al. 2019). It may be attributed to reflective thinking while deciding on the necessary qualities of polymers. The sample's decreased fire retardant might be due to uniformity of about 7.528% (Nguyen, Thu, and Bui 2021; Amin, El-Gamal, and Hashem 2015; Araby et al. 2021; Yu and Lau 2017; Liu et al. 2018). The reason for this reduction is due to the brittleness between the reinforcement and matrix that was not addressed by the organic surfactant.

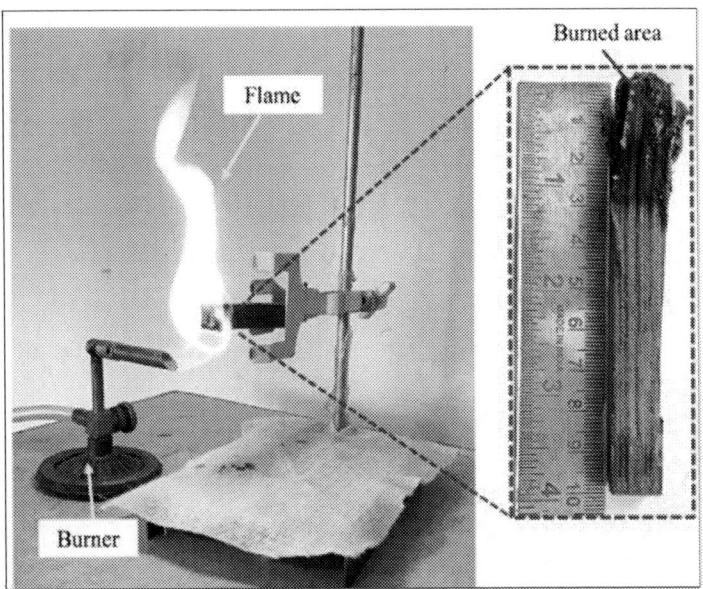

Figure 4.8. Fire retardant test setup.

Summary

The present section is based on the characterization of recycled waste carpets to develop structural composite material. Carpet industries are expanding daily due to the enhancement of their demand; as a result, the generation of waste carpets is also growing simultaneously. The usage of carpets and rugs expanded to several different purposes, *viz.*, for decoration in malls, commercial buildings, and floor covering in houses, hospitals, airports, etc. Waste management of the carpet is the biggest challenging job because the average life span of carpets is 4-6 years. Beyond this duration, it begins to deteriorate and becomes a solid waste. Waste carpet creates environmental pollution after burning or landfilling and creating smells in the environment, etc.

This section is based on the characterization of recycled waste carpet for the development of structural composites. The carpet industry is expanding day by day due to increasing demand. At the same time, the amount of discarded carpet is increasing. The use of carpets and rugs has expanded to several different purposes, such as decoration in shopping malls and commercial buildings, flooring in homes, hospitals, airports, etc. After that period of 4-6 years, it begins to degrade and become solid waste. Discarded carpets cause environmental pollution after incineration or landfill and emit odors into the environment.

This chapter discusses the need to characterize and analyze the properties for demanding applications in order to use composites developed from carpet waste polymers. An overall conclusion based on the knowledge gained can be drawn as follows.

- Thermogravimetric analysis (TGA) has been performed on the fabricated samples to analyze the thermal stability of the developed composite.
- The crystallographic structure and chemical content of the developed composite were determined utilizing X-ray diffraction (XRD) on fabricated samples.
- The interaction between the epoxy and carpet materials was revealed using Fourier transform infrared (FTIR) spectroscopy, which improved the thermomechanical characteristics of the composites.

The outcomes revealed a cost-effective and viable solution for carpet waste management and composite structural development hazardous to and the amount of waste generated in the carpet sector is large and dangerous to the environment and people. Limited data on the recycling of carpet waste, cost-effectiveness, and durability of products are available. Therefore, it is necessary to recycle waste products with the goal of cost-effectiveness and durability of the products. The proposed book could be a new direction for managing discarded carpet waste.

References

Aghaee, Kamran, and Mohammad Foroughi. 2013. "Mechanical Properties of Lightweight Concrete Partition with a Core of Textile Waste." *Advances in Civil Engineering* 2013: 482310. doi:10.1155/2013/482310.

Amin, M.S., S.M.A. El-Gamal, and F.S. Hashem. 2015. "Fire Resistance and Mechanical Properties of Carbon Nanotubes – Clay Bricks Wastes (Homra) Composites Cement." *Construction and Building Materials* 98: 237–49. doi:10.1016/j.conbuildmat.2015.08.074.

Araby, Sherif, Brock Philips, Qingshi Meng, Jun Ma, Tahar Laoui, and Chun H. Wang. 2021. "Recent Advances in Carbon-Based Nanomaterials for Flame Retardant Polymers and Composites." *Composites Part B* 212: 108675. doi:10.1016/j.compositesb.2021.108675.

Awal, A.S.M. Abdul, and Hossein Mohammadhosseini. 2016. "Green Concrete Production Incorporating Waste Carpet Fiber and Palm Oil Fuel Ash." *Journal of Cleaner Production* 137: 157–66. doi:10.1016/j.jclepro.2016.06.162.

Hajsman, Jan, Stepan Jenicek, Ludmila Kucerova, and David Rieger. 2019. "Microstructure and Properties of Polymer Composites." *Manufacturing Technology* 19 (6): 941–46. doi:10.21062/ujep/400.2019/a/1213-2489/mt/19/6/941.

He, Runqin, Qiuxiang Chang, Xinjun Huang, and Jin Bo. 2018. "Improved Mechanical Properties of Carbon FIber Reinforced PTFE Composites by Growing Graphene Oxide on Carbon FIber Surface." *Composite Interfaces* 25 (11): 995–1004. doi:10.1080/09276440.2018.1451677.

Jaiswal, Balram, Rajesh Kumar Verma, and Sanjay Mishra. 2022. "Use of Discarded Carpet Material in the Development of Polymer (Epoxy) Composites for Structural Functions." *The Journal of The Textile Institute*, 1–11. doi:10.1080/00405000.2021.2025302.

Javidparvar, Ali Asghar, Reza Naderi, and Bahram Ramezanzadeh. 2019. "Epoxy-Polyamide Nanocomposite Coating with Graphene Oxide as Cerium Nanocontainer Generating Effective Dual Active/Barrier Corrosion Protection." *Composites Part B: Engineering* 172: 363–75. doi:10.1016/j.compositesb.2019.05.055.

Karaduman, Y., and L. Onal. 2011. "Water Absorption Behavior of Carpet Waste Jute-Reinforced Polymer Composites." *Journal of Composite Materials* 45 (15): 1559–71. doi:10.1177/0021998310385021.

Kumar, Jogendra, Kaushlendra Kuldeep Kaushlendra Kumar, Balram Jaiswal, Kaushlendra Kuldeep Kaushlendra Kumar, and Rajesh Kumar Verma. 2022. "Investigation on the Physio-Mechanical Properties of Carpet Waste Polymer Composites Incorporated with Multi-Wall Carbon Nanotube (MWCNT)." *Journal of the Textile Institute*, 1–10. doi:10.1080/00405000.2022.2062860.

Kumar, Jogendra, Kuldeep Kaushlendra Kumar, Kuldeep Kaushlendra Kumar, Balram Jaiswal, and Rajesh K. Verma. 2022. "Development of Waste Carpet (Jute) and Multi-Wall Carbon Nanotube Incorporated Epoxy Composites for Lightweight Applications." *Progress in Rubber Plastics and Recycling Technology* 38 (3): 247–63. doi:10.1177/14777606221110252.

Kumar, Jogendra, and Rajesh Kumar Verma. 2021. "Experimental Investigation for Machinability Aspects of Graphene Oxide/Carbon Fiber Reinforced Polymer Nanocomposites and Predictive Modeling Using Hybrid Approach." *Defence Technology* 17 (5): 1671–86. doi:10.1016/j.dt.2020.09.009.

Kumar Verma, Rajesh, Balram Jaiswal, Rahul Vishwakarma, Kuldeep Kaushlendra Kumar, Kuldeep Kaushlendra Kumar, Rajesh Kumar Verma, Balram Jaiswal, Rahul

Vishwakarma, and Kuldeep Kaushlendra Kumar. 2021. "Polymer Composite Developed from Discarded Carpet for Light Weight Structural Applications: Development and Mechanical Analysis." *E3S Web of Conferences* 309: 01154 (1-5). doi:10.1051/e3sconf/202130901154.

Liu, Zhi-qi, Zhi Li, Yun-xian Yang, Yan-ling Zhang, Xin Wen, Na Li, and Can Fu. 2018. "A Geometry Effect of Carbon Nanomaterials on Flame Retardancy and Mechanical Properties of Ethylene-Vinyl Acetate/Magnesium Hydroxide Composites." *Polymers* 10: 1028 (1-14). doi:10.3390/polym10091028.

Mishra, Kunal, and Ranji K. Vaidyanathan. 2019. "Application of Recycled Carpet Composite as a Potential Noise Barrier in Infrastructure Applications." *Recycling* 4 (1): 9 (1-11). doi:10.3390/recycling4010009.

Mohammadhosseini, Hossein, Mahmood Md. Tahir, and M.I. Sayyed. 2018. "Strength and Transport Properties of Concrete Composites Incorporating Waste Carpet Fibres and Palm Oil Fuel Ash." *Journal of Building Engineering* 20: 156–65. doi:10.1016/j.jobe.2018.07.013.

Nguyen, Tuan Anh, Thi Thu, and Trang Bui. 2021. "Effects of Hybrid Graphene Oxide with Multiwalled Carbon Nanotubes and Nanoclay on the Mechanical Properties and Fire Resistance of Epoxy Nanocomposite." *Journal of Nanomaterials* 2021: 2862426 (1-7). doi:10.1155/2021/2862426.

Ning, Huiming, Jinhua Li, Ning Hu, Cheng Yan, Yaolu Liu, Liangke Wu, Feng Liu, and Jianyu Zhang. 2015. "Interlaminar Mechanical Properties of Carbon Fiber Reinforced Plastic Laminates Modified with Graphene Oxide Interleaf." *Carbon* 91: 224–33. doi:10.1016/j.carbon.2015.04.054.

Pakravan, Hamid R., and F. Memarian. 2015. "Needlefelt Carpet Waste as Lightweight Aggregate for Polymer Concrete Composite." *Journal of Industrial Textiles* 46 (3): 833–51. doi:10.1177/1528083715598657.

Pathak, Abhishek K., Hema Garg, Mandeep Singh, T. Yokozeki, and Sanjay R. Dhakate. 2019. "Enhanced Interfacial Properties of Graphene Oxide Incorporated Carbon Fiber Reinforced Epoxy Nanocomposite: A Systematic Thermal Properties Investigation." *Journal of Polymer Research* 26 (2): 23 (1-23). doi:10.1007/s10965-018-1668-2.

Tohidi, Shafagh D., Ana Maria, Nadya V. Dencheva, and Zlatan Denchev. 2018. "Microstructural-Mechanical Properties Relationship in Single Polymer Laminate Composites Based on Polyamide 6." *Composites Part B* 153 (July): 315–24. doi:10.1016/j.compositesb.2018.08.106.

Tong, Liyong, Xiannian Sun, and Ping Tan. 2008. "Effect of Long Multi-Walled Carbon Nanotubes on Delamination Toughness of Laminated Composites." *Journal of Composite Materials* 42 (1): 5–23. doi:10.1177/0021998307086186.

Verma, Rajesh Kumar, Balram Jaiswal, Rahul Vishwakarma, Kuldeep Kaushlendra Kuldeep Kaushlendra Kumar, and Kuldeep Kaushlendra Kuldeep Kaushlendra Kumar. 2022. "Water Absorption Study and Characterization of Polymer Composites Developed from Discarded Nylon Carpet." *IOP Conference Series: Materials Science and Engineering* 1228 (1): 012008. doi:10.1088/1757-899x/1228/1/012008.

Wang, Yazhen, Chenglong Wang, Shaobo Dong, Liwu Zu, and Tianyu Lan. 2020. "The Study on Microstructure and Mechanical Properties of Multi-Component Composite

Based on HDPE." *Designed Monomers and Polymers* 23 (1): 164–76. doi:10.1080/15685551.2020.1818956.

Yu, Zechuan, and Denvid Lau. 2017. "Evaluation on Mechanical Enhancement and Fire Resistance of Carbon Nanotube (CNT) Reinforced Concrete." *Coupled Systems Mechanics* 6 (3): 335–49. doi:10.12989/csm.2017.6.3.335.

Zhang, R. L., B. Gao, W. T. Du, J. Zhang, H. Z. Cui, L. Liu, Q. H. Ma, C. G. Wang, and F. H. Li. 2016. "Enhanced Mechanical Properties of Multiscale Carbon Fiber/Epoxy Composites by Fiber Surface Treatment with Graphene Oxide/Polyhedral Oligomeric Silsesquioxane." *Composites Part A: Applied Science and Manufacturing* 84: 455–63. doi:10.1016/j.compositesa.2016.02.021.

Chapter 5

Applications of Discarded Carpets Composites

Carpets are broadly used as floor coverings, textile-based materials for comfort and insulation in offices, commercial complexes, and domestic materials. In our daily lives, almost all of us see carpets in different forms in our surroundings. On a global level, people use a staggering amount of carpet each year, resulting in enormous waste (Henckens 2021). Carpet waste of about 40,000 tons is sent to a landfill in the UK itself (Mohammadhosseini et al. 2018). Therefore, carpet waste is now a primary environmental and economic concern due to the costs associated with disposal. Compared to the amount of carpet waste produced worldwide each year, only a small portion is recycled (Goswami 2009; Jain et al. 2012). Due to this, it has become a global concern. Carpet waste recycling requires multiple processing steps, making it difficult and expensive (Realff, Ammons, and Newton 1999). Burning this fibrous waste releases highly toxic fumes, which harm human health (Mishra et al. 2019; Ghobakhloo and Fathi 2021; J. Wang et al. 2021). It is a type of solid waste that is causing complex environmental and economic issues. Carpet waste can be collected in two forms, depending on its source of origin (Ahmed et al. 2021; Teli 2018). Thus, it becomes essential to find some solution to convert such solid waste into a functional form with several applications. This chapter highlights the fundamental development product from recycled carpet and textile products. A novel approach is presented to utilize carpet waste to generate lightweight structural applications. In addition, it will reduce the discarded carpet amount and a simultaneously environmentally friendly approach shall be helpful for the manufacturing sector.

Introduction to Polymer-Based Composites

In the 21st century, waste has become the prime concern for the manufacturing sector and industries. Waste management is critical for controlling many types of waste, including solid, liquid, and gaseous forms. Numerous techniques are used in industry, research organizations, and institutions to control waste and resource development (Wilson 2007). The manufacturing of carpets and their uses is a substantial source of waste. In this context, the many techniques employed for reusing carpet waste products. The manufacturing sector generates waste from the initial product development to the end products. The generation of waste leads to an imbalance of environmental conditions and hazardous effects on the ecology. The proper utilization and recycling of discarded products is a feasible solution for waste management (Islam and Bhat 2019; Siddique, Khatib, and Kaur 2008). The principle of "waste to wealth" could be helpful for manufacturing industries and human beings. Due to their bulky sizes and decomposition cost, the waste generated from the carpet and textile industries is a matter of concern. Exploration for the production of discarded carpet polymer materials is in its early stages and requires greater interest from academia, research, manufacturing, and organizations.

Polymers have been extensively used in the manufacturing sector for three decades to meet the varying demands of customers and industries. Nowadays, polymer material efficiently substitutes traditional engineering materials due to exceptional features like reduced weight, corrosion resistance, and improved mechanical aspects. Composites made by using polymers and modified polymers provide several benefits. They differ from traditional production materials because they offer many distinctive qualities, such as reduced density, better strength, processing ease, cost-effectiveness, etc. However, thermoplastic composites have frequently failed to meet consumer requirements for physical strength in industrial applications. In this case, adding fiber increased the mechanical strength of polymer-based composite materials. In this context, research into the manufacture of laminated polymer composite materials is still in its early stages and needs increased interest from academic, scientific, industrial, and institutional sources.

It is typically for the following reasons that composites are chosen over conventional materials such as metal alloys or wood (Hung 2006). Table 5.1 lists the benefits of composite materials over traditional materials.

The main application area of polymer composite material is practical in several fields, including transportation, defense, vehicle, aerospace, sporting goods, and building infrastructure (A. Kumar, Sharma, and Dixit 2019; Hanemann and Szabó 2010; Javidparvar, Naderi, and Ramezanzadeh 2019; Hazra and Basu 2016). Certainly, it is easier to adapt, has a high strength-to-weight ratio, has excellent resistance to moisture, and is a user-friendly design, the material has a variety of applications. The discarded fiber polymer is light, strong, stiff, durable, impact-resistant, moisture-resistant, and chemically stabilized. As a result, it is commonly employed in producing automotive and construction equipment and materials (Müller et al. 2017). In the series, discarded carpet epoxy composites such as roadside barriers, wall tiles, and roof tiles are used in noise reduction. It has recently been used in tooling component development, pumps, and drive components (chain driver, belt) (M. K. A. Ali, Hou, and Abdelkareem 2020; Nguyen Bich and Nguyen Van 2016). It has also been used to manufacture naval components such as masts and long straight upright (deck) poles made of laminated polymeric materials (Jiang, Qi, and Qin 2019; Jin et al. 2021).

Table 5.1. Advantages of composite material

Sr. No.	Physical significance	Remark
1	Cost	High mass production, Part consolidation, Long term durability and Less Manufacture time
2	Weight	Low weight load distribution and high strength-to-weight ratio
3	Dimension	Tailored geometry
4	Surface properties	Corrosion and moisture resistance.
5	Thermal properties	Less thermal conductivity.
6	Electrical properties	High dielectric strength, Non-magnetic, and Radar transparency

In the following points, the application of polymeric laminates is summarized herein.

- Aerospace Industry, Aeronautics, Automobiles, and Aircraft wings industries (Long et al. 2019)
- Electronics Packaging, electronics application (Phiri, Gane, and Maloney 2017)
- Electromagnetic interface shielding (Li, Zhang, and Zhang 2017)
- Fuel cells and their Bipolar Plates (Rzeczkowski, Krause, and Pötschke 2019)
- Wind turbine blades (Pathak et al. 2016)
- Sporting goods, rackets, and golf sticks (Choudhary and Gupta 2001)
- Bio-medical (Choudhary and Gupta 2001)
- Academic area (Lee et al. 2015)

Light Weight Structure Tooling Component

The goal of composite structures is to reduce the overall weight of the product while reducing mass to maintain or improve product functionality. Less dense materials, such as metal foams and composites, or reduced volume of material should be used in structurally important components. Under both conditions, less energy is required to deliver the finished product, supporting the eco-friendly components in lightweight construction. For lightweight composite structures, fiber-reinforced thermoplastic materials could be used in manufacturing. The main purpose of using lightweight materials is weight reduction and possible cost savings. Significant weight savings and improved performance reduce fuel usage and greenhouse gas emissions. Reducing weight and stress concentrations is a key goal for performance optimization, to ensure the required level of performance.

When it comes to automotive applications, polymeric materials have become a popular choice because of their superior performance and ability to be tailored to specific needs. These materials are manufactured to provide superior strength, corrosion resistance to most toxins, and high durability in the harshest environmental conditions while being relatively lightweight. Polymer materials offer significant advantages over conventional metallic materials in terms of fatigue and fracture resistance, 20–40% weight savings, fast process cycles, ability to meet stringent dimensional stability requirements, and low thermal expansion properties. In contrast, in the automotive industry, engineering polymers are always needed to replace low-cost, widely accessible, and easily processed materials like polypropylene, polyethylene, polyamides, etc. The vehicle's aesthetics are an additional advantage. Material recyclability gets enhanced in this scenario because of the large percentage of goods that can be reused. Recently, energy-saving lightweight materials were developed from materials that are 95% recyclable, 10% was recovered by energy recovery or thermal recycling and 85% was recovered by reuse or mechanical recycling (Sandin and Peters 2018). By eliminating the liner requirement, vehicle engine dimensions can be significantly reduced. For a typical car weighing 1100 kg, a weight saving of about 110 kg results in a 7% increase in fuel efficiency (Lamberti, Román-Ramírez, and Wood 2020; Fontaras, Zacharof, and Ciuffo 2017). Weight reduction, manufacturing techniques and recycling are key to reducing $CO2$ in the transport sector, especially ships, cars, trucks, and vans. Light weight is very important as functional operations account for about 85% of energy consumption over the entire life cycle. There are many ways to reduce CO_2

emissions, including weight reduction. The most effective way is to reduce the amount of CO_2 emitted by any energy source. Many systems are carefully combined to create entirely new structural or functional characteristics that cannot be addressed by individual components. Due to its lightweight, it is corrosion resistant, durable, flexible mounting and durable. They are widely used in automotive structures, electronic packaging, medical equipment construct, and the construction of housing and other structures.

Advanced composites incorporate high-performance fiber reinforcements into a polymer matrix such as epoxy. Epoxy composites include Kevlar/epoxy and glass/carbon. Some commercial applications for advanced composites were previously reserved for the building & construction area. Particulate composites, fibrous composites, and laminate composites are all subcategories based on the type of reinforcement used (consisting of laminates). They can be further segmented using non-biodegradable and biodegradable matrices. The term "green composites" refers to bio-based composites constructed from natural/biofiber and biodegradable polymers. Hybrid and textile composites can be further segmented. These composite materials have a long history and prominent applications in building construction. They have been used in low-volume applications due to their shorter lead times and cheaper investment costs than conventional materials. Polymer composites have gained popularity due to the lightweight and integration potential of components as well as design flexibility, corrosion resistance, material anisotropy, and tensile strength.

Despite these advantages, the use of polymer composite has been hampered by high material costs, slow production rates, and minor concerns about product recyclability. In spite of their success in the high-performance industry, polymer composites have not yet been widely applied in large-scale automotive mass production. These include concerns about the material's ability to absorb and manage crash energy, recycling difficulties competitive and economic constraints. The industry also lacks overall expertise and comfort with this material.

Fiber-reinforced thermoplastic composites are becoming more popular in engineering applications, while thermosets dominate the automotive market. Resin solutions are commonly used in the commercial production of composites. Various polymers are available depending on the basic raw material. Within each general category are several subcategories.

The most commonly used polymers are polyester, vinyl ester, epoxy, phenolic, polyimide, polyamide, polypropylene, polyether ether ketone, and others. While crushed minerals and fibers can be used as reinforcements.

Vacuum infusion results in a product with 40% resin and 60% percent fiber, whereas hand layup produces a product with 60% resin and 40% fiber content. This ratio has a significant impact on the strength of the product. Carpet waste PMCs are replaceable with polymer material because of their low cost and ease of production. Mechanical properties, such as strength, modulus, and impact resistance, limit the usage of nonreinforced polymers in structures. The use of recycled carpet waste fibers with strong fiber networks to reinforce the polymer enables the manufacture of PMCs.

Composite materials such as fiber and high-performance resins have helped modern vehicles and overcome the problems posed by their sophisticated designs. Advanced composite materials now play a key role in the manufacture of military vehicles, solar panels, etc. (Mangalgiri 1999). Composite polymers have boomed in the automotive sector over the last 40 years to create lightweight and durable parts. Initially, small parts, such as car bodies, were produced, but the technology has since expanded to larger and more critical parts, such as engine components. Due to the significant weight savings over metallic designs, polymer composite applications are being used in a variety of manufacturing platforms. Laminate composites are used in various vessels, including sailboats, kayaks, canoes, speedboats, luxury yachts, navy ships, and submarines. Due to their low cost, composites made from recycled carpet waste polymer composites can be used to construct over 95% of naval vehicle manufacturing (Oladele et al. 2022; Youjiang Wang 1999). Fiber composites are used in other high-performance applications. It is also used in the construction of boat hulls, masts, bulkheads, deck houses, and seating, as well as cable ladders-trays, rudders, and ventilation ducts.

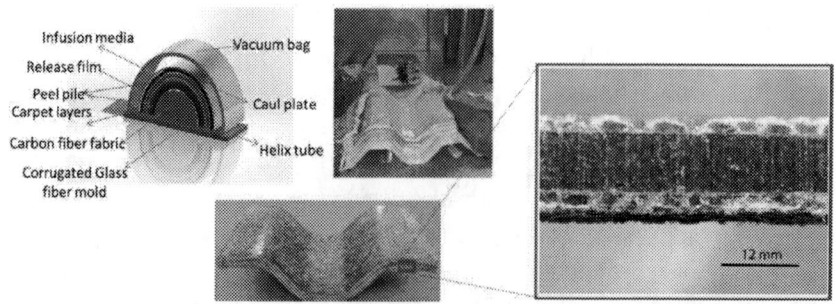

Figure 5.1. Carpet waste polymer composites for tooling component (Mishra et al. 2019).

Nowadays, tooling materials use as polymer composite materials. These materials are also found in critical parts, such as frames, gear panels, and tooling. In order to increase the rigidity/stiffness, aluminum alloy is added to the waste carpet material to increase its strength. It is also used in various tooling applications, such as antennas, solar panels, and naval (Ship) materials. Figure 5.1 demonstrates the application of tooling components and novel applications.

Sound Insulation Materials

The carpet materials are commonly used in floor coverings and home products. However, because of its inability to degrade, it is polluting the environment. Because of this, there is a pressing need to eliminate waste. Toxic vapors are released into the air when solid trash is burned, while recycling is not an option due to contamination. As a result of these factors, recycling into new items is the better option (Mishra et al. 2019). Hand layup (J. Kumar, Abhishek, and Xu 2022; Rajak et al. 2019) and VARTM (J. Kumar, Kumar, Kumar, et al. 2022; Jain et al. 2012) are two methods for recycling carpets into polymer composites. Mechanical recycling, which involves collecting, shredding, and pelletizing garbage before reusing it in producing new products, is an appealing option for disposing of it. Composite materials that withstand loads can be made from recycled carpet reinforced with epoxy. Carpet waste fiber-reinforced polymer composites (CFRPC) are gaining popularity due to their unique features, economic efficiency, ease of collection, and reduced environmental impact. In addition, because they are a renewable resource, they offer a more long-term supply solution (Halliwell 2006). In addition to this, sound pollution can be controlled in the building industry, roadside barriers, and movie theaters (Kumar Verma et al. 2021). It is possible to use carpet waste because of the variety of fibers that can be used as reinforcements from appliance by-products (Ailenei et al. 2021).

Wall/Roof Tiles

Researchers have been looking to manage the solid waste carpet fiber that can be used to reinforce composite materials and has similar physical and mechanical qualities (Sotayo, Green, and Turvey 2015). This is because fiber-reinforced polymers are expensive and have an adverse effect on the environment. Despite the benefits, fiber is naturally hydrophilic and has a higher moisture absorption rate, making it difficult to fabricate composites. The fiber's mechanical characteristics are negatively impacted by the swelling and softening caused by moisture absorption, and the hydrophilic nature of the moisture has an impact on how it disperses and mixes within the matrix phase. For these reasons, the fiber must be altered physically to improve its interaction with the polymer matrix.

Ali et al. (Azam Ali et al. 2015; A. Ali et al. 2017) studied how jute fiber reinforcement affects the mechanical properties and moisture recovery of composites. They found that the treated jute fiber composites recovered a lower moisture content and exhibited better mechanical characteristics. Kumar et al. (J. Kumar, Kumar, Jaiswal, et al. 2022) evaluated the water absorption/swelling effect of carpet waste (jute fiber) composites. They found that the composites improved their moisture resistance. Bakri et al. (Jayamani et al. 2014) studied the mechanical, morphological, and spectral properties of banana fiber/epoxy composite and evaluated the impacts of alkaline treatment. The treated fiber composites were found to have superior properties. The fiber-like kenaf, oil palm, bamboo, jute, sisal, coconut, and pineapple leaf fiber in natural reinforced composites has been the subject of numerous studies. Carpet fiber can be used as a fiber source derived from industrial waste. These waste fibers have been matrices with both thermoset and thermoplastic resins. Polyester, polypropylene, and wool reinforcement material are widely used in the production of composite materials. Kumar et al. (J. Kumar, Kumar, Kumar, et al. 2022) have studied jute-epoxy composites.

They manufactured composites using the VARTM method and evaluated the effects of jute fiber on the morphological, chemical, mechanical, and thermal properties of the composites. They found that the mixing process of jute fiber was applicable in Tile/ruff tile. In a similar study, Verma et al. (Verma et al. 2022) also confirmed that polymer composites developed from nylon waste carpet can be used to fabricate tiles. In addition, a numerical stud conducted on drying carpet tile confirmed the sustainable temperature and moisture ability (Francis and Wepfer 1993). Similar results indicated that the successful addition of treated PALF as a reinforcement to LDPE could successfully produce useful composites with good properties. The increased fiber fraction improved the mechanical properties of the produced/manufactured composites, demonstrating the reinforcement potential of PALF. Therefore there is a possibility of industrial use of this fiber waste. For example, the production of bathroom wall tiles (as shown in Figure

5.2). The characteristics of the developed tiles were determined to be as satisfactory as mentioned in the graph (Figure 5.3).

Figure 5.2. Recycle polymer composites tiles (Gebremedhin and Rotich 2020).

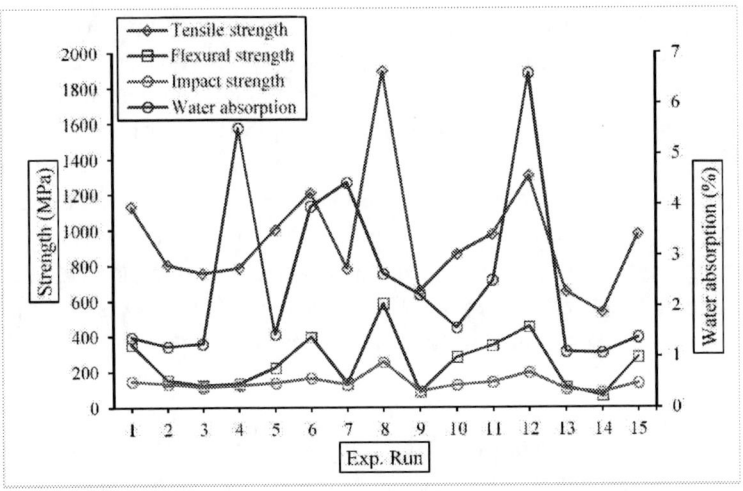

Figure 5.3. Properties of developed composites tile (Gebremedhin and Rotich 2020).

Roadside Barriers

As industrial wastes pile up, they pose a severe environmental and public health hazard. Eminent scholars have been looking for ways to recycle carpet waste that can be utilized to strengthen composite materials and create products with similar sound-insulating properties. This is because the sound-absorbing, multilayered synthetic fiber has a good influence on the environment. Considering their and advantages, fiber is highly sound absorbing and ideal for use in composites. The ability of the fiber to absorb sound is influenced by different fiber materials, with environmental absorption properties affecting and coalescing within the matrix phase. For these reasons, carpet waste fibers have been modified to improve interaction with the polymer matrix. Carpet waste accounts for about 44 million tons of waste in the UK, producing a considerable number of by-products. The technical qualities/properties of carpet waste enable its reuse to be reasonable and easy from an environmental and economic perspective. Many authors have explored the possibility of using carpet waste instead of gravel in product manufacturing. Noise pollution is a major problem that is gaining importance due to its detrimental effects on human health. Most of the outdoor noise pollution comes from transportation systems such as car and rail noise.

The World Health Organization (WHO) estimates that Europe loses at least one million healthy lives yearly due to disability and disease caused by traffic noise. Under the European directive on the Assessment and Management of Environmental Noise, new programs have been established, which are expected to encourage an increase in the usage of noise barriers to lessen traffic noise. To minimize sound reflection towards noise-sensitive locations next to the highway, conventional barriers are typically designed to reflect a significant amount of traffic noise. Due to the requirement for noise-absorbing traffic noise barriers, significant advancements in the field of sound-absorbing materials have been made. A composite of carpet waste composites is one of the most popular materials used for highway noise barriers (Rushforth et al. 2005). By using threads and yarn aggregate, carpet fiber is formed. Thus, inherent friction inside the yarn of the material causes it to absorb sound energy (Miraftab, Horrocks, and Woods 1999). Therefore, the study provides a dual benefit to the environment. On the one hand, industrial waste is recycled, and on the other, road noise is reduced. The thickness of the panel and the mix of several layers with fiber have been impacted to achieve the product's largest acoustic insulating behavior. It was determined that the finished composite product might be used as an outdoor building element, and an environmental assessment was conducted that included leachability and radioactivity tests. Recycled carpet, fiberglass, rubber, and wood fiber appear to be the most viable materials for use in roadside safety applications. Because concrete has such high strength, developing these goods to match them makes sense. It is, therefore possible to apply recycled material to numerous applications where concrete is currently employed. All of these uses warrant further examination, including guardrail posts and offset blocks, sign supports, sign blanks, and barricades. Flexible delineator posts, channelizing drums, and traffic cones are just a few examples of other possible uses.

Recycled materials are used economically in roadside safety applications. As long as recycled materials can be proven to outlast conventional materials in terms of life expectancy, they will be more competitively priced. Increased demand for these products should lead to lower manufacturing costs and a more competitive pricing structure due to the use of high-volume production lines. Those who use recycled material may have had unpleasant experiences with recycled products in the past. This must be overcome. Creep and flexural difficulties are being addressed by manufacturers, who are strengthening their goods to mitigate the inherent weaknesses of most plastics. Differences in the thermal expansion of plastics and other materials also change construction methods and connection details. As consumer awareness grows and recycling efforts deepen, the quality and purity of post-consumer waste streams will improve.

Seven applications of carpet waste composites are developed due to commercially available goods. Barriers, bollards, delineators (including flexible posts, drums, and cones), guardrail offset blocks, posts, sign blanks, and sign supports are all included in this category. Most manufacturers believe that their products are competitive with conventional materials when the product's lifespan is considered.

Summary

Engineered qualities particular to a given application will necessitate the use of composite polymer materials for engineering applications in the future. The development of composite in tools components for an application require high performance. From the mid-1990s to the present day, polymer materials have undergone a period of rapid and promising development. As a result, specialized polymer materials for use in and outside of transportation vehicles have been created, serving a wide range of purposes and industries. This means that polymer materials have become dominant in various sectors, including printing, electronics, civil engineering, and transportation. This way, the "utilization cycle" is completed, allowing further developments in other industries and back into transport applications. Compared to conventional materials, composites may suit a wide range of design needs with significant weight reductions and high strength-to-weight ratios.

Composite materials provide the following advantages over conven-tional ones:

- Composites can be tailored to meet performance and sound-absorbing requirements to reduce noise and sound pollution.
- Water absorption and swelling are excellent properties of carpet waste polymer composites. In addition to having smoother surfaces, composites can easily incorporate integral decorative wall tiles/roof tiles.
- Composite parts can eliminate joints and fasteners, allowing for part simplification and integrated design in comparison to conventional materials.

Despite their potential advantages in terms of weight reduction and corrosion resistance, it seems unlikely that advanced composites will find widespread use in automotive applications.

Modern composites need extensive and drastic changes for their wide range of commercial applications in automobiles and trucks. Raw and manufactured materials are more expensive than current alternatives. Advanced composites, on the other hand, have been used for specialized components in the commercial vehicle sector.

Advanced composite materials have the opportunity to demonstrate their performance benefits in specialty vehicles of a variety of types, which are produced in limited numbers. The composite industry worldwide is using recycled carpet waste in thermosets to improve the molding of polymer composites. Composites used in automobiles account for 24% of the global market for thermoset composites. Carpet fiber/epoxy composites for rocker arm covers, suspension arm housings, engine components, and filament-wound fuel tanks are likely future automotive opportunities. Fibers of thermoset polymers reinforced from carpet waste are relatively inexpensive, and the material's fast cycle time and capacity allow for the integration of parts, making it a potentially lightweight material.

Another possibility is fiber-reinforced polymers made from carpet waste, but their use in sound insulation/absorption requires significant improve-ments in manufacturing cost and efficiency. Advanced composites are a viable option for the wall/roof tiles industry and the need to reduce sound pollution.

References

Ahmed, Hemn Unis, Rabar H. Faraj, Nahla Hilal, Azad A. Mohammed, and Aryan Far H. Sherwani. 2021. "Use of Recycled Fibers in Concrete Composites: A Systematic Comprehensive Review." *Composites Part B: Engineering* 215: 108769. doi:10.1016/j.compositesb.2021.108769.

Ailenei, Eugen Constantin, Savin Dorin Ionesi, Ionut Dulgheriu, Maria Carmen Loghin, Dorina Nicolina Isopescu, Sebastian George Maxineasa, and Ioana Roxana Baciu. 2021. "New Waste-Based Composite Material for Construction Applications." *Materials* 14 (20): 1–14. doi:10.3390/ma14206079.

Ali, A., V. Baheti, A. Jabbar, J. Militky, S. Palanisamy, H. F. Siddique, and D. Karthik. 2017. "Effect of Jute Fibre Treatment on Moisture Regain and Mechanical Performance of Composite Materials." *IOP Conference Series: Materials Science and Engineering* 254 (4): 042001. doi:10.1088/1757-899X/254/4/042001.

Ali, Azam, Khubab Shaker, Yasir Nawab, Munir Ashraf, Abdul Basit, Salma Shahid, and Muhammad Umair. 2015. "Impact of Hydrophobic Treatment of Jute on Moisture Regain and Mechanical Properties of Composite Material." *Journal of Reinforced Plastics and Composites* 34 (24): 2059–68. doi:10.1177/0731684415610007.

Ali, Mohamed Kamal Ahmed, Xianjun Hou, and Mohamed A.A. Abdelkareem. 2020. "Anti-Wear Properties Evaluation of Frictional Sliding Interfaces in Automobile Engines Lubricated by Copper/Graphene Nanolubricants." *Friction* 8 (5): 905–16. doi:10.1007/s40544-019-0308-0.

Choudhary, Veena, and Anju Gupta. 2001. "Polymer/Carbon Nanotube Nanocompo-sites." In *Carbon Nanotubes - Polymer Nanocomposites*, edited by Siva Yellampalli, 65–90. Intech Open. doi:10.5772/18423.

Fontaras, Georgios, Nikiforos Georgios Zacharof, and Biagio Ciuffo. 2017. "Fuel Consumption and CO2 Emissions from Passenger Cars in Europe – Laboratory versus Real-World Emissions." *Progress in Energy and Combustion Science* 60. Elsevier Ltd: 97–131. doi:10.1016/j.pecs.2016.12.004.

Francis, Nicholas D., and William J. Wepfer. 1993. "Experimental and Numerical Analysis of the Drying Characteristics of Modular Carpet Tiles." *Textile Research Journal* 63 (1): 1–13. doi:10.1177/004051759306300101.

Gebremedhin, Negasi, and Gideon K. Rotich. 2020. "Manufacturing of Bathroom Wall Tile Composites from Recycled Low-Density Polyethylene Reinforced with Pineapple Leaf Fiber." *International Journal of Polymer Science* 2020: 2732571. doi:10.1155/2020/2732571.

Ghobakhloo, Morteza, and Masood Fathi. 2021. "Industry 4.0 and Opportunities for Energy Sustainability." *Journal of Cleaner Production* 295: 126427. doi:10.1016/j.jclepro.2021.126427.

Goswami, K. K. 2009. *Advances in Carpet Manufacture*. Edited by K.K. Goswami. *Advances in Carpet Manufacture*. 2nd ed. United Kingdom: Elsevier B.V. doi:10.1533/9781845695859.

Halliwell, Sue. 2006. "End of Life Options for Composite Waste Recycle, Reuse or Dispose ? National Composites Network Best Practice Guide." *National Composites Network*. https://compositesuk.co.uk/system/files/documents/endoflifeoptions.pdf.

Hanemann, Thomas, and Dorothée Vinga Szabó. 2010. "Polymer-Nanoparticle Composites: From Synthesis to Modern Applications." *Materials* 3 (6): 3468–3517. doi:10.3390/ma3063468.

Hazra, Surajit, and Sukumar Basu. 2016. "Graphene-Oxide Nano Composites for Chemical Sensor Applications." *C — Journal of Carbon Research* 2 (2): 12. doi:10.3390/c2020012.

Henckens, Theo. 2021. "Scarce Mineral Resources: Extraction, Consumption and Limits of Sustainability." *Resources, Conservation and Recycling* 169: 105511. doi:10.1016/j.resconrec.2021.105511.

Hung, Ching-Cheh. 2006. Carbon Materials Metal/Metal Oxide Nanoparticle Composite and Battery Anode Composed of the Same. US 7094499 B1, issued 2006.

Islam, Shafiqul, and Gajanan Bhat. 2019. "Environmentally-Friendly Thermal and Acoustic Insulation Materials from Recycled Textiles." *Journal of Environmental Management* 251: 109536. doi:10.1016/j.jenvman.2019.109536.

Jain, Abhishek, Gajendra Pandey, Abhishek K. Singh, Vasudevan Rajagopalan, Ranji Vaidyanathan, and Raman P. Singh. 2012. "Fabrication of Structural Composites from Waste Carpet." *Advances in Polymer Technology* 31 (4): 380–89. doi:10.1002/adv.20261.

Javidparvar, Ali Asghar, Reza Naderi, and Bahram Ramezanzadeh. 2019. "Epoxy-Polyamide Nanocomposite Coating with Graphene Oxide as Cerium Nanocontainer Generating Effective Dual Active/Barrier Corrosion Protection." *Composites Part B: Engineering* 172: 363–75. doi:10.1016/j.compositesb.2019.05.055.

Jayamani, Elammaran, Sinin Hamdan, Md Rezaur Rahman, Soon Kok Heng, and Muhammad Khusairy Bin Bakri. 2014. "Processing and Characterization of Epoxy/Luffa Composites: Investigation on Chemical Treatment of Fibers on Mechanical and Acoustical Properties." *BioResources* 9 (3): 5542–56. doi:10.15376/biores.9.3.5542-5556.

Jiang, Tianjia, Longbin Qi, and Wei Qin. 2019. "Improving the Environmental Compatibility of Marine Sensors by Surface Functionalization with Graphene Oxide." Research-article. *Analytical Chemistry* 91 (20). American Chemical Society: 13268–74. doi:10.1021/acs.analchem.9b03974.

Jin, Huichao, Limei Tian, Wei Bing, Jie Zhao, and Luquan Ren. 2021. "Toward the Application of Graphene for Combating Marine Biofouling." *Advanced Sustainable Systems* 5 (1): 1–16. doi:10.1002/adsu.202000076.

Kumar, Amit, Kamal Sharma, and Amit Rai Dixit. 2019. "A Review of the Mechanical and Thermal Properties of Graphene and Its Hybrid Polymer Nanocomposites for Structural Applications." *Journal of Materials Science* 54 (8): 5992–6026. doi:10.1007/s10853-018-03244-3.

Kumar, Jogendra, Kumar Abhishek, and Jinyang Xu. 2022. "Experimental Investigation on Machine-Induced Damages during the Milling Test of Graphene/Carbon Incorporated Thermoset Polymer Nanocomposites." *Journal of Composites Science* 6 (77): 1–12. doi:10.3390/jcs6030077.

Kumar, Jogendra, Kaushlendra Kuldeep Kaushlendra Kumar, Balram Jaiswal, Kaushlendra Kuldeep Kaushlendra Kumar, and Rajesh Kumar Verma. 2022. "Investigation on the Physio-Mechanical Properties of Carpet Waste Polymer Composites Incorporated with Multi-Wall Carbon Nanotube (MWCNT)." *Journal of the Textile Institute*, 1–10. doi:10.1080/00405000.2022.2062860.

Kumar, Jogendra, Kuldeep Kaushlendra Kumar, Kuldeep Kaushlendra Kumar, Balram Jaiswal, and Rajesh K. Verma. 2022. "Development of Waste Carpet (Jute) and Multi-Wall Carbon Nanotube Incorporated Epoxy Composites for Lightweight Applications." *Progress in Rubber Plastics and Recycling Technology* 38 (3): 247–63. doi:10.1177/14777606221110252.

Kumar Verma, Rajesh, Balram Jaiswal, Rahul Vishwakarma, Kuldeep Kaushlendra Kumar, Kuldeep Kaushlendra Kumar, Rajesh Kumar Verma, Balram Jaiswal, Rahul Vishwakarma, and Kuldeep Kaushlendra Kumar. 2021. "Polymer Composite Developed from Discarded Carpet for Light Weight Structural Applications: Development and Mechanical Analysis." *E3S Web of Conferences* 309: 01154 (1-5). doi:10.1051/e3sconf/202130901154.

Lamberti, Fabio M., Luis A. Román-Ramírez, and Joseph Wood. 2020. "Recycling of Bioplastics: Routes and Benefits." *Journal of Polymers and the Environment* 28 (10): 2551–71. doi:10.1007/s10924-020-01795-8.

Lee, Miyeon, Jihoon Lee, Sung Young Park, Byunggak Min, Bongsoo Kim, and Insik In. 2015. "Production of Graphene Oxide from Pitch-Based Carbon Fiber." *Scientific Reports* 5: 1–10. doi:10.1038/srep11707.

Li, An, Cong Zhang, and Yang Fei Zhang. 2017. "Thermal Conductivity of Graphene-Polymer Composites: Mechanisms, Properties, and Applications." *Polymers* 9 (9): 437 (1-17). doi:10.3390/polym9090437.

Long, Jia Peng, San Xi Li, Bing Liang, and Zhi Guo Wang. 2019. "Investigation of Thermal Behaviour and Mechanical Property of the Functionalised Graphene Oxide/Epoxy Resin Nanocomposites." *Plastics, Rubber and Composites* 48 (3): 127–36. doi:10.1080/14658011.2019.1577027.

Mangalgiri, P. D. 1999. "Composite Materials for Aerospace Applications." *Bull. Mater. Sci* 22 (3): 657–64. doi:10.1088/1755-1315/166/1/012006.

Miraftab, Mohsen, Richard Horrocks, and Colin Woods. 1999. "Carpet Waste, an Expensive Luxury We Must Do Without." *Autex Research Journal* 1 (1): 1–7. doi:10.1533/9780857092991.3.173.

Mishra, Kunal, Sarat Das, Ranji Vaidyanathan, and Tooling Materials. 2019. "The Use of Recycled Carpet in Low-Cost Composite Tooling Materials." *Recycling* 4: 12 (1-8). doi:10.3390/recycling4010012.

Mohammadhosseini, Hossein, Mahmood Tahir, Abdul Rahman, Mohd Sam, Nor Hasanah, Abdul Shukor, and Mostafa Samadi. 2018. "Enhanced Performance for Aggressive Environments of Green Concrete Composites Reinforced with Waste Carpet Fibers and Palm Oil Fuel Ash." *Journal of Cleaner Production* 185 (1): 252–65. doi:10.1016/j.jclepro.2018.03.051.

Müller, Kerstin, Elodie Bugnicourt, Marcos Latorre, Maria Jorda, Yolanda Echegoyen Sanz, José M. Lagaron, Oliver Miesbauer, Alvise Bianchin, Steve Hankin, Uwe Bolz, German Perez, Marius Jesdinszki, Martina Linder, Zuzana Scheuerer, Sara Castello and Markus Schmid. 2017. "Review on the Processing and Properties of Polymer Nanocomposites and Nanocoatings and Their Applications in the Packaging,

Automotive and Solar Energy Fields." *Nanomaterials* 7 (4): 74. doi:10.3390/nano7040074.

Nguyen Bich, Ha, and Hieu Nguyen Van. 2016. "Promising Applications of Graphene and Graphene-Based Nanostructures." *Advances in Natural Sciences: Nanoscience and Nanotechnology* 7 (2): 023002. doi:10.1088/2043-6262/7/2/023002.

Oladele, Isiaka Oluwole, Samson Oluwagbenga Adelani, Okikiola Ganiu Agbabiaka, and Miracle Hope Adegun. 2022. "Applications and Disposal of Polymers and Polymer Composites : A Review." *Euro. J. Adv. Engg. Tech.* 9 (3): 65–89.

Pathak, Abhishek K., Munu Borah, Ashish Gupta, T. Yokozeki, and Sanjay R. Dhakate. 2016. "Improved Mechanical Properties of Carbon Fiber/Graphene Oxide-Epoxy Hybrid Composites." *Composites Science and Technology* 135: 28–38. doi:10.1016/j.compscitech.2016.09.007.

Phiri, Josphat, Patrick Gane, and Thad C. Maloney. 2017. "General Overview of Graphene: Production, Properties and Application in Polymer Composites." *Materials Science and Engineering B* 215: 9–28. doi:10.1016/j.mseb.2016.10.004.

Rajak, Dipen Kumar, Durgesh D. Pagar, Pradeep L. Menezes, and Emanoil Linul. 2019. "Fiber-Reinforced Polymer Composites: Manufacturing, Properties, and Applications." *Polymers* 11 (10): 1667 (1-37). doi:10.3390/polym11101667.

Realff, Matthew J., Jane C. Ammons, and David Newton. 1999. "Carpet Recycling: Determining the Reverse Production System Design." *Polymer-Plastics Technology and Engineering* 38 (3): 547–67. doi:10.1080/03602559909351599.

Rushforth, I. M., K. V. Horoshenkov, M. Miraftab, and M. J. Swift. 2005. "Impact Sound Insulation and Viscoelastic Properties of Underlay Manufactured from Recycled Carpet Waste." *Applied Acoustics* 66: 731–49. doi:10.1016/j.apacoust.2004.10.005.

Rzeczkowski, Piotr, Beate Krause, and Petra Pötschke. 2019. "Characterization of Highly Filled PP/Graphite Composites for Adhesive Joining in Fuel Cell Applications." *Polymers* 11 (3): 462. doi:10.3390/polym11030462.

Sandin, Gustav, and Greg M. Peters. 2018. "Environmental Impact of Textile Reuse and Recycling – A Review." *Journal of Cleaner Production* 184: 353–65. doi:10.1016/j.jclepro.2018.02.266.

Siddique, Rafat, Jamal Khatib, and Inderpreet Kaur. 2008. "Use of Recycled Plastic in Concrete: A Review." *Waste Management* 28 (10): 1835–52. doi:10.1016/j.wasman.2007.09.011.

Sotayo, Adeayo, Sarah Green, and Geoffrey Turvey. 2015. "Carpet Recycling: A Review of Recycled Carpets for Structural Composites." *Environmental Technology and Innovation* 3: 97–107. doi:10.1016/j.eti.2015.02.004.

Teli, M. D. 2018. "Finishing of Carpets for Value Addition." In *Advances in Carpet Manufacture*, edited by K. K. Goswami, Second Edi, 175–211. United Kingdom: Elsevier Ltd. doi:10.1016/B978-0-08-101131-7.00010-1.

Verma, Rajesh Kumar, Balram Jaiswal, Rahul Vishwakarma, Kuldeep Kaushlendra Kuldeep Kaushlendra Kumar, and Kuldeep Kaushlendra Kuldeep Kaushlendra Kumar. 2022. "Water Absorption Study and Characterization of Polymer Composites Developed from Discarded Nylon Carpet." *IOP Conference Series: Materials Science and Engineering* 1228 (1): 012008. doi:10.1088/1757-899x/1228/1/012008.

Wang, Junxiong, Jiakuan Yang, Huijie Hou, Wei Li, Jingping Hu, Mingyang Li, Wenhao Yu, Zhongyi Wang, Sha Liang, Keke Xiao, Bingcchuan Liu, Kai Xi, R. Vasant Kumar. 2021. "A Green Strategy to Synthesize Two-Dimensional Lead Halide Perovskite via Direct Recovery of Spent Lead-Acid Battery." *Resources, Conservation and Recycling* 169 (February): 105463. doi:10.1016/j.resconrec.2021.105463.

Wang, Youjiang. 1999. "Utilization of Recycled Carpet Waste Fibers for Reinforcement of Concrete and Soil." Edited by Y. Wang. *Polymer-Plastics Technology and Engineering* 38 (3). Cambridge, UK: Woodhead Publishing Limited: 533–46. doi:10.1080/03602559909351598.

Wilson, David C. 2007. "Development Drivers for Waste Management." *Waste Management & Research* 25 (3): 198–207. doi:10.1177/0734242X07079149.

Appendix 1

m^2K/W	Unit of Thermal Insulation
gm/cc	Unit of Density
t	Thickness of sample
Wt.%	Weight %
w/w	Weight by weight percentage
v/v	Volume by volume percentage
Hrs	Hours
mPa-s	Unit of Viscosity
MPa	Unit of Tensile/Flexural
°C	Unit of Temperature
FS	Flexural strength
t_m	Machining time
w_i	Initial weight
w_f	Final weight
ρ	Density
%	Percentage
cm^{-1}	Unit of Wave Number

Index

A

acrylic carpets, 8
aluminium oxide (Al$_2$O$_3$), xi
American standard testing materials (ASTM), xii, 51, 52, 55, 56, 57, 60, 61, 78, 79, 84
application, 9, 12, 24, 30, 32, 38, 40, 43, 45, 48, 55, 68, 69, 77, 88, 92, 93, 97, 101, 104, 106
automotive shredder residue (ASR), xii, 29

B

back front-front back, xi, 53, 57
blends carpets, 8

C

carbon-fiber reinforced composites (CFRCs), 47
carpet backing cloth (CBC), xii
carpet waste, v, vii, 1, 10, 11, 12, 14, 15, 17, 18, 19, 21, 27, 31, 32, 34, 35, 37, 38, 41, 42, 43, 45, 47, 48, 50, 53, 54, 55, 58, 61, 62, 63, 64, 65, 66, 67, 68, 69, 71, 72, 73, 74, 76, 77, 79, 81, 82, 83, 84, 86, 87, 88, 91, 92, 96, 97, 98, 99, 100, 101, 102, 104, 105, 106, 107
carpet-derived fuel (CDF), xi, 14
ceramic matrix composite (CMC), xi, 23, 25, 40, 41
CF/GF reinforced polymer interlaminar strength, 46
composite(s), v, vii, ix, xi, 1, 12, 15, 16, 17, 18, 19, 21, 22, 23, 24, 25, 26, 27, 30, 31, 32, 33, 34, 35, 36, 37, 38, 39, 40, 41, 42, 43, 44, 45, 46, 47, 48, 50, 51, 53, 54, 55, 57, 58, 59, 60, 61, 62, 63, 64, 65, 66, 67, 68, 69, 71, 72, 73, 74, 75, 76, 77, 78, 79, 80, 81, 82, 83, 84, 85, 86, 87, 88, 89, 91, 92, 93, 94, 95, 96, 97, 98, 99, 100, 101, 102, 103, 104, 105, 106, 115, 116

D

density (ρ), xi, 21, 22, 26, 27, 34, 36, 49, 50, 92, 109

E

epoxy, viii, 17, 23, 24, 32, 33, 34, 35, 36, 38, 40, 41, 42, 43, 45, 47, 51, 54, 57, 58, 59, 60, 61, 63, 65, 66, 67, 68, 69, 70, 71, 72, 73, 74, 75, 76, 77, 79, 80, 81, 82, 83, 84, 86, 87, 88, 89, 93, 95, 97, 98, 102, 104, 105, 106
epoxy composite, viii, 40, 41, 42, 45, 51, 57, 58, 60, 61, 65, 67, 70, 71, 73, 74, 75, 77, 87, 89, 93, 98, 102, 105

F

fiber reinforced concrete (FRC), xii, 28, 49
fiber reinforcement polymer (FRP), xi, 31, 66
fiber type carpets, 8
fiber-reinforced thermoplastic materials, 94
final weight (w_f), 109
fire retardants, 84
flatweave, 5, 7
flatweave carpets, 5, 7
flax, 15, 17, 49, 63, 67
flexural strength (FS), 27, 55, 57, 63, 64
fly ash (FA), xii, 17
Fourier-transform infrared spectroscopy (FTIR), xi, 74, 76, 86

Index

front back-back front (FBBF), xi, 36, 53, 54, 57, 58, 59, 64

G

gaseous, vii, 1, 91
glass fiber reinforcement polymer (GFRP), xi, 28, 43, 46, 50
graphene nanoplate/multiwall carbon nanotube (GNP/MWCNT), xii
graphene/carbon fiber reinforcement polymer (G/CFRP), xi

H

hand-woven carpets, 3, 8
hardener (H), xi, 16, 22, 24, 34, 36, 40, 41, 43, 65, 66, 67, 74, 75, 87, 89, 103
hemp, 49
Hessian cloth (HC), xii
high strength concrete (HSC), xii, 29, 38
hours (Hrs), 36, 61, 78, 79, 109

I

industrial waste, vii, 12, 98, 99, 100
initial weight (w_i), 109
interlaminar flexural shear strength (IFSS), xi, 45
interlaminar shear strength (ILSS), xi, 35, 45

J

jute, 15, 17, 41, 49, 50, 60, 63, 67, 77, 78, 80, 81, 87, 98, 103, 105

L

light weight aggregates (LWA), xii
lightweight composite structures, 94
lightweight materials, 94
liquid, vii, 1, 24, 79, 91
low-density polyethylene (LDPE), xi, 27, 41, 98, 103

M

machine-made, 2, 6, 18
machining time (t_m), 109
maleic anhydride (MAh), xi, 27
metal matrix composite (MMC), xi, 24
morphological characterization, 71

N

natural coarse aggregate (NCA), xii, 29
needle-felt carpets, 8
non-woven carpets, 7
nylon, xi, xii, 2, 8, 15, 18, 24, 27, 34, 38, 48, 49, 50, 52, 53, 54, 56, 57, 58, 59, 63, 64, 68, 73, 76, 77, 83, 84, 88, 98, 106
nylon carpets, 8, 28, 49, 88, 106
nylon fiber fabric reinforced concrete (NFFRC), xii
nylon fiber reinforced concrete (NFRC), xii
nylon fiber-based composites (NFC), xi
nylon reinforced composites (NRC), xi

O

olefin carpets, 8
olefin fiber-based composite (OFC), xi

P

palm oil fuel ash (POFA), xi, 81
percentage (%), 10, 33, 94, 109
poly vinyl chloride (PVC), xi, 27, 50
polyester, 2, 8, 24, 33, 34, 38, 40, 48, 50, 52, 53, 54, 56, 57, 58, 59, 63, 64, 67, 72, 73, 74, 76, 77, 81, 83, 95, 98
polyester carpets, 8
polymer matrix composite (PMC), xi, 24, 33, 44
polymer-based composites, 91
polypropylene (PP), xi, 2, 7, 12, 15, 17, 18, 28, 30, 33, 34, 39, 41, 48, 50, 52, 53, 54, 56, 57, 58, 59, 63, 64, 65, 67, 68, 73, 76, 81, 83, 94, 95, 98, 106

polyvinyl alcohol (PVA), xi
post-consumer waste, 11, 101

R

recycled polyethylene terephthalate (rPET, xi, 27
recycled polyethylene terephthalate (rPET), xi, 27
recycled waste ceramic aggregate (RWCA), xi, 29, 38
recycling, iii, ix, 1, 10, 12, 13, 14, 15, 16, 17, 18, 27, 29, 30, 32, 38, 41, 42, 44, 48, 50, 67, 68, 77, 78, 81, 86, 87, 88, 91, 92, 94, 95, 97, 101, 104, 105, 106, 107, 116
resin (R), viii, xi, 1, 24, 28, 30, 32, 33, 34, 35, 36, 38, 42, 47, 49, 63, 75, 78, 82, 84, 95, 96, 105
resin transfer, viii, xi, 28, 34, 49, 78
roadside barriers, 77, 93, 97, 99

S

ship wool fiber (SWF), xii
silicon carbide (SiC), xi, 25
sound insulation, 4, 27, 34, 43, 97, 102, 106
spectroscopy emission method (SEM), xi, 28, 61, 62

T

tensile test, 51, 52, 64
thermogravimetric analysis (TGA), xii, 82, 83, 84, 86
tufted carpet, 7, 18

U

unconfined compressive strength (UCS), xii

unit of density (gm/cc), 109
unit of tensile/flexural (MPa), 30, 34, 53, 54, 59, 63, 64, 109
unit of thermal insulation (m^2K/W), 9, 109
unit of viscosity (mPa-s), 109
unit of wave number (cm^{-1}), 74, 75, 76, 109
universal testing machine (UTM), xii, 51, 52, 55, 56

V

vacuum assisted resin transfer moulding (VARTM), xi, 15, 30, 34, 35, 36, 37, 38, 39, 40, 49, 50, 63, 78, 97, 98
velvet carpets, 8
volume by volume percentage (v/v), 109

W

wall/roof tiles, 97, 102
waste, vii, xi, xii, 1, 10, 11, 12, 14, 15, 16, 17, 18, 19, 21, 27, 30, 31, 32, 33, 34, 35, 36, 37, 38, 39, 40, 41, 43, 44, 45, 46, 47, 48, 50, 51, 54, 57, 59, 61, 63, 64, 65, 66, 67, 68, 69, 71, 72, 75, 77, 79, 81, 82, 83, 84, 85, 86, 87, 88, 91, 96, 97, 98, 100, 102, 103, 104, 105, 106, 107, 115, 116
waste carpet fiber (WCF), xii, 18, 28, 33, 38, 39, 46, 49, 65, 68, 87, 97, 105
weight % (Wt.%), 109
weight by weight percentage (w/w), 109
wool, xii, 2, 8, 27, 33, 34, 50, 52, 53, 54, 56, 57, 58, 59, 64, 73, 75, 76, 83, 98
wool carpets, 8
woven carpet, 3, 7

X

X-ray diffraction analysis (XRD), xi, 72, 73, 77, 86

About the Authors

Prof. Dr. Rajesh Kumar Verma is a Professor in the Department of Mechanical Engineering, School of Engineering at the Harcourt Butler Technical University, Kanpur, India. It is a renowned and old premier institute whose roots are as deep as its outlook. "Government Research Institute, Cawnpore" was established in 1920 and was renamed "Government Technological Institute" in 1921. Finally, in 1926 it got the name we know today, "Harcourt Butler Technological Institute." As per Act No. 11 of 2016 by the Government of Uttar Pradesh, it has become a university, i.e., Harcourt Butler Technical University Kanpur, India. Prof. Verma received his PhD (Engineering) from Jadavpur University, Kolkata, India. He is actively involved in teaching and research in waste management, fiber science, composites, nanomaterial, modeling, simulation, optimization, and Manufacturing processes. Dr. Verma currently supervised/ongoing more than 13 Masters and 09 PhD thesis and published more than 103 research articles in peer-reviewed journals and conferences. The e-content and courses he developed are widely used in the Indian university system. He has completed/ongoing more than eight nos. of research and development (R&D) projects sponsored by various govt. agencies. He is now a reviewer of various international publishers such as Elsevier, sage, Springer, etc. Moreover, Dr. Verma has been invited as Keynote/Invited Speaker, Section Chair, and member for several international conferences with themes in Mechanical & Production Engineering, Emerging Materials, and manufacturing science. He has developed various nanomaterials for structural applications and proposed hybrid optimization modules recognized by multiple peer-reviewed journal publishers like Elsevier, Springer, Taylor and Francis, Sage, etc.

E-mail: rkvme@hbtu.ac.in

Dr. Jogendra Kumar is currently working as an Assistant Professor in NIET, NIMS School of Mechanical and Aerospace Engineering, NIMS University Rajasthan, Jaipur, India. He has completed a doctorate (PhD) degree from the Madan Mohan Malaviya University, Gorakhpur, India, in 2021. He has been working in developing composite materials, characterization, and machining for the last 4 years. Recently, he has completed a Govt. sponsored R&D

Project based on polymer composites development from discarded fibers. His research focuses on recycling waste materials, Nanomaterial synthesis and development, Nanocomposite's machining, numerical modeling, and optimization. He has published his work in international peer-reviewed journals and book chapters.

E-mail: jogendra.kumar@nimsuniversity.org

Er. Shivi Kesarwani is currently a Senior Research Assistant (SRA) in the Department of Mechanical Engineering at the Madan Mohan Malaviya University of Technology, India. His research interests focus on nanomaterial synthesis and development, composites machining, numerical modeling, and surface texturing. Additionally, he has utilized various nanomaterials reinforced fibrous composites for structural applications. He has developed hybrid metaheuristic optimizing methodologies recognized by multiple peer-reviewed journals and book publishers like Elsevier, Springer, AIP, etc.

E-mail: skme@mmmut.ac.in